AF621716

# Wealth Creation without Pollution

# Wealth Creation without Pollution

*Designing for Industry, Ecobusiness Parks and Industrial Estates*

Edited by
Brian D'Arcy, Lee-Hyung Kim and Marla Maniquiz-Redillas

IWA PUBLISHING

**Published by** **IWA Publishing**
**Alliance House**
**12 Caxton Street**
**London SW1H 0QS, UK**
Telephone: +44 (0)20 7654 5500
Fax: +44 (0)20 7654 5555
Email: publications@iwap.co.uk
Web: www.iwapublishing.com

First published 2018
© 2018 IWA Publishing

Apart from any fair dealing for the purposes of research or private study, or criticism or review, as permitted under the UK Copyright, Designs and Patents Act (1998), no part of this publication may be reproduced, stored or transmitted in any form or by any means, without the prior permission in writing of the publisher, or, in the case of photographic reproduction, in accordance with the terms of licenses issued by the Copyright Licensing Agency in the UK, or in accordance with the terms of licenses issued by the appropriate reproduction rights organization outside the UK. Enquiries concerning reproduction outside the terms stated here should be sent to IWA Publishing at the address printed above.

The publisher makes no representation, express or implied, with regard to the accuracy of the information contained in this book and cannot accept any legal responsibility or liability for errors or omissions that may be made.

**Disclaimer**
The information provided and the opinions given in this publication are not necessarily those of IWA and should not be acted upon without independent consideration and professional advice. IWA and the Editors and Author will not accept responsibility for any loss or damage suffered by any person acting or refraining from acting upon any material contained in this publication.

*British Library Cataloguing in Publication Data*
A CIP catalogue record for this book is available from the British Library

ISBN: 9781780408330 (paperback)
ISBN: 9781780408347 (eBook)
ISBN: 9781780408804 (ePub)

**Cover image:** www.istockphoto.com

# Contents

# Primum non nocere
## (First do no harm)

### Primum non nocere needs to be more than just a mantra if we are to avoid leaving a legacy of environmental destruction for future generations.

Pollution is primarily anthropogenic and is the result of humans not thinking about the bigger picture. To quote the book preface 'pollution is inefficiency and loss of resources'. It's not just about wasting valuable resources but we will be responsible for the long-term damaging impact on the environment and essential ecosystems if we continue on this path.

We believe that the key to eradicating pollution is to engage with the broad range of influencers and decision makers throughout the water sector. This book provides an excellent overview of current best practice solutions from across the world and we hope that it will encourage the exchange of ideas to enable sustainability leadership.

As an organisation, we use our expertise to improve the built environment and maintain the quality of the natural and cultural environments. Not seeking to destroy existing habitats and respecting the 'sense of place' enjoyed by local communities, we endeavour to prevent or treat pollution by closely evaluating projects and working with our clients to deliver a more sustainable outcome.

We are continuing to push the boundaries of stormwater management in urban areas in order to work towards a truly sustainable future. We believe that together we can shape a better world, but there is a long journey ahead. Publications like this provide a roadmap to help us get there.

**About Arup**

Arup people are driven to discover new ways to turn ideas into tangible reality. This passion is behind many of the world's most prominent projects in the built environment and across industry. We offer a broad range of professional services that combine to make a real difference to our clients and the communities in which we work.

www.arup.com

ARUP

The Korean Society on Water Environment, established in 1985, is an academic and professional institution aiming to promote academic and technology development in the field of water environment related to water quality control and drainage wherein almost 2,300 members are currently involved.

KSWE focuses on solutions regarding the serious deterioration of water quality affecting human lifestyle due to rapid industrialization and urbanization. Several research studies on technology and policies about the various water quality and aquatic environment problems contribute on the establishment of Korea's own version of environmental management systems. Paradigm shifts in the water environment management were requisites to climate change and continuous watershed and land use alteration. The KSWE has boosted its efforts to change from passive into individual management of pollution sources and integration of infrastructures for water quality management leading to a change of paradigm. Nevertheless, the institution works on developing exceptional and future-oriented water management technologies such as the convergence of information technology and biotechnology into water mitigation strategies, integrated water management approach and formulation of several policies. In addition, through the active participation of members, the KSWE strives to expand globally through conferences inviting professionals from international academic societies.

The *Journal of Korean Society on Water Environment* is the official journal publication of the KSWE. It was first amended as the 'Journal of Korean Society of Water Quality' in June 1998 and was renamed again into 'Journal of Korean Society on Water Quality' with pISSN 122-4144 in June 1999. Currently, the journal is known as the 'Korean Society on Water Environment' which was changed in January 2012. The pISSN number was changed into '2289-0971' with eISSN of 2289-098X in May 2014. The manuscripts accepted in the journal are classified into Research papers, Review Papers, Technical Notes, Special Issues, Discussions, etc. concerning water quality control wherein the papers are accepted written either in Korean or English. The topics included in the journal are as follows: hydraulic & hydrologic, soil & groundwater, aquatic ecosystem, water & sewage, nonpoint pollution, water quality modeling, watershed management, environmental analysis, environmental health & toxicology, environmental policy, and climatic environment.

The journal is published six times a year during the 30th day of January, March, May, July, September, and November. Through the years, the journal has published 33 volumes with 2,056 research papers up until January 2017.

**www.kswe.org**

Korean Society on Water Environment

# Sponsors

The following organisations have helped in the production of this book in various ways, not least by holding meetings, seminars, and conferences which featured issues and speakers to progress the information gathering and authorship encouragement needed for the book.

C & D Associates LLP, UK
Chartered Institution of Water and Environmental Management, CIWEM, (Scottish Branch) UK
De La Salle University-Manila, Philippines
Diffuse Pollution & Eutrophication Specialist Group of the International Water Association (IWA)
Environmental Engineering association, Thailand (EEAT)
Green Industry Network (GIN)
Industrial Estates Authority, Thailand (IEAT)
Kongju National University, Cheonan, Korea
Melbourne Water, Melbourne, Australia
Research Centre for Environmental and Hazardous Substance Management, Khon Kaen University, Thailand
Tuil Solutions, Scotland UK
University of Abertay Dundee, UK

# List of Chemical Symbols and Standard Acronyms

| | |
|---|---|
| ABS | Acrylonitrile butadiene styrene |
| Al | Aluminium |
| As | Arsenic |
| B | Boron |
| Ba | Barium |
| BHC | Benzene hexachloride |
| $BOD_5$ | Biochemical oxygen demand (5 indicates a standard 5-day duration measurement) |
| C | Carbon |
| Ca | Calcium |
| Cd | Cadmium |
| COD | Chemical oxygen demand |
| Cr | Chromium |
| Cr+6 | Hexavalent chromium (most toxic form) |
| COD | Chemical oxygen demand |
| Chl-a | Chlorophyll-a |
| Cl | Chlorine |
| Cl-H/C | Chlorinated hydrocarbons |
| CN | Cyanide |
| Cu | Copper |
| DDT | Dichlorodiphenyltrichloroethane |
| DO | Dissolved oxygen |
| DTN | Dissolved total nitrogen |
| DTP | Dissolved total phosphorus |
| EC | Electrical conductivity |
| F | Fluorine |
| Fe | Iron |

| | |
|---|---|
| FIOs | Faecal indicator organisms |
| H | Hydrogen |
| HC | Hydrocarbons |
| HCH | Hexachlorocyclohexane (also known as Lindane; an isomer of BHC) |
| Hg | Mercury |
| $H_2O$ | Water |
| I | Iodine |
| K | Potassium |
| Li | Lithium |
| Mg | Magnesium |
| Mn | Manganese |
| N | Nitrogen |
| Na | Sodium |
| $NH_3$ | Ammonia |
| $NH_4$-N | Ammonium (strictly $NH^{4+}$) |
| Ni | Nickel |
| $NO_3$ | Nitrate |
| O | Oxygen |
| P | Phosphorus |
| Pb | Lead |
| pH | A measure of the concentration of hydrogen ions, and hence the acidity or alkalinity of a solution |
| PAH | Polyaromatic hydrocarbon |
| PCBs | Polychlorinated biphenyl hydrocarbons |
| POP | Priority organic pollutants |
| $PO_4$-P | Phosphate |
| S | Sulphur |
| $SO_2$ | Sulphur dioxide |
| $SO_4$-S | Sulphate |
| Sn | Tin |
| SS | Suspended solids |
| TBT | Tributyl tin |
| Ti | Titanium |
| TN | Total nitrogen |
| TOC | Total organic carbon |
| TP | Total phosphorus |
| TSS | Total suspended solids |
| TOC | Total organic carbon |
| U | Uranium |
| Zn | Zinc |

## GENERAL GUIDE TO TECHNICAL ACRONYMS

| | |
|---|---|
| BAT | Best available techniques |
| BMPs | Best management practices |
| BREFs | Best available techniques reference documents |
| CETP | Common effluent treatment plant |
| CSO | Combined sewer overflow |
| DOC | Dissolved organic carbon |
| DWF | Dry weather flow |
| EC | European Commission |
| ETP | Effluent treatment plant |
| EIP | Eco-industrial park |
| EMC | Event mean concentration |
| EIAs | Environmental impact assessments |
| EQS | Environmental quality standard (typically a target concentration of a pollutant) |
| EU | European Union |
| FEH | Flood estimation handbook, produced by Institute of Hydrology (UK) |
| FWS | Free-water surface |
| GI | Green infrastructure |
| GIS | Geographic information system |
| GRV | Groundwater recharge volume |
| HSSF | Horizontal subsurface flow |
| IED | Industrial Emissions Directive |
| IMP | Integrated management practice |
| IMPEL | EU network for Implementation and Enforcement of Environmental Law |
| IPPC | Integrated Pollution Prevention and Control (from European Community (EC) directive 2008/1/EC on integrated pollution prevention and control). |
| LID | Low impact development (drainage aspects) |
| PAH | Polycyclic aromatic hydrocarbons |
| POP | Priority organic pollutants |
| RBMP | River Basin Management Plan |
| STW | Sewage treatment works |
| SWO | Surface water outfall (drainage pipe discharge point) |
| SUDS | Sustainable urban drainage systems |
| TMDL | Total maximum daily load |
| TN | Total nitrogen |
| TON | Total oxidised nitrogen |

| | |
|---|---|
| TPH | Total petroleum hydrocarbons |
| TP | Total phosphorus |
| TSS | Total suspended solids |
| VSSF | Vertical subsurface flow |
| WFD | Water Framework Directive |
| WRAP | Winter rainfall acceptance potential |
| WRAP | Waste and Resources Action Programme (UK) |
| WQV | Water quality volume |
| UWWTD | Urban Waste Water Treatment Directive |
| WSUD | Water sensitive urban design |
| WWTP | Waste water treatment plant |

# Preface

This publication aims to bring to the attention of the next generations of industrial researchers and practitioners, whether organic chemists, engineers, trainee managers and economists, or current plant managers and site operators, facts about water pollution associated with industry and industrial products. The book should also help environmental managers in non-governmental organisations, public agencies and local authorities, water utilities and developers, architects and planners. The book is also written for policymakers and political leaders and their advisors, to help better understand and address pollution issues. Pollution is inefficiency and loss of resources. It is a cost to the polluter, and a cost in loss of water resources to other businesses as well as to local people and their health and quality of their environment.

We are pleased to have achieved part of those aims by securing the high-level endorsement by political leaders in the two countries where the editors have been based during the production of the book (UK and South Korea), with parallel Forewords from the Right Honourable Sir Vince Cable from the UK, and Dr Younghoon Kim, Director General, Water Environment Policy Bureau, Ministry of Environment, Republic of Korea. We are grateful to both of them for their support. We deliberately chose an old industrial country and a new modern one to span the spectrum of industrial/commercial development. The authors of the papers in the book are from many countries, from Australia to the USA, and Netherlands and Germany to China and Korea. They cite research and case studies from all the developed continents, including studies on polar pollution, tropical issues, and best practice experience in North America and elsewhere.

The book is the output of several years of deliberations by academics, environmental consultants and regulators, engaging with industrial and commercial sectors to characterise and quantify environmental problems and identify best practice solutions. Equally important have been efforts to explore the essential regulatory regimes that can provide a basis of effective means to prevent pollution and achieve good working environments in which industry and commerce can flourish. Of necessity, the content is multidisciplinary, as are the real world issues to be addressed.

The book would not have been possible without the support of the International Water Association (IWA), especially the Diffuse Pollution Specialist Group. We are also grateful for encouragement from the Green Industry Network (GIN). The editors are grateful to all the contributing authors of the constituent papers in the book, and to co-authors and reviewers too. We are also grateful to the sponsors of the book, including the organisers of a succession of conferences and workshops which have generated interest in this project. Finally, we are grateful to Maggie Smith and Mark Hammond and colleagues at IWA Publishing for their patience and help getting this book completed.

Brian J D'Arcy, Lee-Hyung Kim and Marla Maniquiz-Redillas, (Editors) 13th March 2017.

# Foreword from Vince Cable

## *Wealth Creation without Pollution – Designing for Industry, Ecobusiness Parks and Industrial Estates*

Industrialisation has featured in the transformation of economies and societies, generating wealth as well as products, stimulating trade, and developing relationships between source areas of raw materials and consumers of manufactured products. Increasingly, the patterns of economic activity are now more evenly spread as most countries are developing manufacturing capacity. Globalisation favours an economic 'level playing field' for free trade between nations; environmental best practice needs to be the same.

Whilst helping to provide many basic attributes of modern life as well as luxuries, industrial and commercial activities have also caused environmental degradation and pollution. Clean, safe water supplies are a basic commodity, as are clean air and safe, high quality food. When one industry pollutes a river, it is not just an ecological problem. There are economic impacts on communities and other businesses dependent on good water quality. Economic impacts may be widespread if a country or region becomes known as polluted, unsafe or unhealthy.

There are lessons to be learned and shared. There is already a wealth of experience and expertise in developing industrial and commercial practices which do not create pollution problems. Cleaner production processes and intelligent site design are opportunities for all producers in the interests not just of health and environmental protection, but for sustaining viable businesses, without wasting resources or risking accidents, and without constraining others who need clean unpolluted water.

Globalisation has sometimes led to the 'export' of polluting industries to other countries, rather than the export of clean technology and expertise. In a free trade world, it is increasingly difficult for one country to stand alone in developing cleaner technologies and practices which may not always pay back over a short-term period, but still require concerted action in the long-term interests of everyone, including business.

This joint publication by scientists and engineers from many countries is to be welcomed as a contribution to a more sustainable business future for the benefit of all countries, and people everywhere.

Vince Cable<br>The Right Honourable Sir Vince Cable MP

Dr Cable was Secretary of State for Business, Innovation and Skills, and President of the Board of Trade in the UK Government from 12th May 2010–11th May 2015. Dr Cable was Member of Parliament for Twickenham from May 1997 to May 2015, and re-elected on 8th June 2017. An economics graduate from Cambridge University, Dr Cable got his PhD from Glasgow University, and worked as a professional economist in Kenya and then advising the Commonwealth Office before returning to lecture at Glasgow University. He was Chief Economist for Shell from 1995–1997 prior to being elected for the Liberal Democrats as MP for Twickenham in 1997. Dr Cable continues to be an active and respected figure in politics in Britain and internationally.

# Foreword from South Korea

Urbanisation and industrialisation are processes in which the natural ground cover is transformed into an artificial one. During the course of the urbanisation process, the environment becomes polluted and the ecosystem is damaged. Recently, the United Nations established the Sustainable Development Goals (SDGs) with key elements including social development, economic growth, and environmental protection aimed to solve economic inequality on a global level, environmental degradation, and the severely damaged ecosystem. 'Green industry' has been adopted as a worthy solution to achieve environmental protection and economic growth based on an eco-friendly system that runs resource and energy efficiently, producing low carbon and waste emission, via a non-polluting and safe industry through integrated lifecycle management. Industrial ecology for an Eco-Industrial Park (EIP), which resembles the mass circulation and energy flow of nature, is focused on potential roles of businesses in reducing environmental pollutant loads with management of the lifecycle of product manufacture. Ecological engineering is a sound technological approach for solving the various environmental problems in the industrial complex. Recently, various fields such as landscaping, civil engineering, environmental design, policy, and energy technology have been grafted with ecological engineering in a wide range of industrial complexes.

This book, comprising of five chapters, dealing with the industrial impacts on the water environment, the eco-business parks concept, sustainable drainage systems for industry and commerce, environmental regulations, and methods on how to improve existing pollution problems, could provide important knowledge on the environmental pollution problems, technical and regulatory solutions, and examples of stormwater management applied in industrial sites.

I hope that this book serves as a tool in establishing new policies and developing more advanced techniques for the management of stormwater pollution in industries, eco-business parks and industrial estates. I recommend this book to regulators, policymakers, planners, developers, and practitioners engaging with industrial and commercial sectors as well as to academics, researchers and students in environmental science and engineering fields, and to those who aim to create a wealthy environment without pollution.

**Younghoon Kim**
Director General
*Water Environment Policy Bureau*
*Ministry of Environment, Republic of Korea*

# Chapter 1

# Industrial pollution and the water environment: a historical perspective

*B. J. D'Arcy*, L.-H. Kim and Peter Morrison*

**Corresponding author: brian@enviroexperience.co.uk*

## 1.1 INTRODUCTION

In the UK and much of Western Europe, the traditional image of smoke stack industries polluting the air, with oily and toxic effluent streams ruining rivers and coasts, is increasingly a historical one. As the home of the Industrial Revolution, many of the rivers and estuaries of the United Kingdom (Britain) were severely impacted by industrial waste streams. Heavily coloured with dyes, process effluents from textiles had high biological oxygen demand (BOD; also a characteristic of food industry and paper making effluents), whilst toxic metals and other pollutants were characteristic of tanneries, engineering, metal finishing and associated industries. Kay (1832) described the environmental conditions in Manchester during the Industrial Revolution:

- '*The* (River) *Irk, black with the refuse of dye-works erected on its banks, receives* … (drainage from) … *the gas works, and filth of the most pernicious character from bone-works, tanneries, size manufacturers,* etc.
- (There is) *no common slaughter house in Manchester, and those which exist are chiefly in* (the) *narrowest and most filthy streets in the town. The drainage from these houses, deeply tinged with blood, and impregnated with other animal matters, frequently flows down the common surface drain of the street* …'

Similar conditions prevailed in the other industrialising cities, prior to development of proper sewer systems and treatment works, and before modern techniques for recovering value from waste and adequately treating trade effluents

© IWA Publishing 2017. Wealth Creation without Pollution: Designing for Industry, Ecobusiness Parks and Industrial Estates
Brian D'Arcy, Lee-Hyung Kim, Marla Maniquiz-Redillas
doi: 10.2166/9781780408330_01

were available. In the UK and Germany a whole spectrum of organic pollutants were discharged from the developing chemical industry, which – whilst revolutionising technology and products available for humanity – destroyed many miles of watercourses. Porter (1973) reviewed the impact of industry in four industrialised estuaries in Britain, making the case for government actions to reduce the pollution problems. Table 1.1 gives a quantitative idea of the contribution of local industries to the polluted condition of one of those estuaries, the Mersey.

**Table 1.1** The main industrial pollution loads discharging to the Mersey Estuary, UK, in 1971, prior to the Control of Pollution Act, 1974.

| Type of Industry | No. of Dischargers | BOD Pollution Load in kg/day |
|---|---|---|
| Inorganic chemicals, metals | 20 | 12,725 |
| Organic chemicals | 5 | 11,550 |
| Mineral oil refining & petrochemicals | 4 | 42,850 |
| Detergents, vegetable oil refining | 3 | 4841 |
| Food industries (excluding vegetable oil refining) | 4 | 1331 |
| Animal waste processing | 3 | 6625 |
| Paper mills | 2 | 12,693 |
| **Total** | **41** | **92,615** |

*Source:* Modified from Porter (1973).

A decade later and great improvements had been made on the Mersey (D'Arcy, 1988), and by the mid-1990s the impacts of industrial effluent discharges were in decline in many of the initially industrialising countries. In the UK, for example, in 1995 the Forth River Purification Board (FRPB) reported that only about 10% of its polluted waters were caused by industrial effluent discharges to the freshwater reaches of the river system (FRPB, 1995). Although the basis of the reporting system was changed on implementation of the European Water Framework Directive, to give a broader indication of good ecological status, the adverse impact of industry in relation to process effluent discharges, has continued to decline in Scotland.

Table 1.2 lists the relative importance of industrial impacts compared with other water pollution sources currently in the USA (Environmental Protection Agency [EPA], 2014). Evidence for industrial pollution impacts is reported in the USA as impairments – actual or threatened impairment of potential use. Only a proportion of States reported data to the USEPA, so the figures do not necessarily represent a national picture. For assessed rivers and streams, industry ranked 12th as a pollution cause. When the miles of impairment due to industry are expressed as a percentage of impaired rivers and streams, industry accounts for only 2.2% of total reported impairments.

In many industrialised countries there was a shift of industry to the coast, to areas closer for supply of raw materials and for export of products. Some large industries perhaps also sought locations where large volumes of difficult effluent could be discharged into a perceived greater degree of dilution. For bays and estuaries in the USA (Table 1.3), the continuing importance of industrial discharges is still evident, although still causing fewer impairments than atmospheric deposition, 'unknown', and municipal discharges/sewage. Although the 4th largest individual cause, industry accounted for only 9% of the total impaired area for bays and estuaries in the USA (USEPA accessed 14.3.2014, in http://iaspub.epa.gov/waters10/attains_nation_cy.control#total_assessed_waters).

**Table 1.2** National summary of probable sources of impairments in assessed rivers and streams in the USA.

| Ranking | Probable Source Category | Miles Threatened or Impaired |
|---|---|---|
| 1 | Agriculture | 125,180 |
| 2 | Unknown | 101,903 |
| 3 | Atmospheric deposition | 99,622 |
| 4 | Hydromodification | 57,997 |
| 5 | Urban-related runoff/stormwater | 56,068 |
| 6 | Municipal discharges/sewage | 53,860 |
| 7 | Natural/wildlife | 52,365 |
| 8 | Unspecified nonpoint source | 48,270 |
| 9 | Habitat alterations (not directly related to hydromodification) | 33,732 |
| 10 | Resource extraction | 28,835 |
| 11 | Silviculture (Forestry) | 19,558 |
| ***12*** | ***Industrial*** | ***16,022*** |
| 13 | Construction | 12,668 |
| 14 | Other | 9231 |
| 15 | Land application/waste sites/tanks | 8168 |
| 16–22 | Sum of additional 7 minor categories, including legacy pollutants, aquaculture, recreation, military bases, etc. | 10,115 |

*Source:* USEPA accessed March 14, 2014, in http://iaspub.epa.gov/waters10/attains_nation_cy.control#total_assessed_waters.

The decline in industrial impacts has been a result of three factors:

(a) Economic decline and failure of old industries to modernise and be more efficient; such businesses were often serious polluters and many have closed.
(b) Improvements in effluent quality occurred as the economic value of materials comprising the effluent was recognised; leading to more efficient use of resources, product or raw material recovery, and waste minimisation philosophy and practices.
(c) Development of better treatment technology, as a process focus was applied across a business driven by regulatory requirements in parallel with the business need in (b) above.

**Table 1.3** National summary (USA): Probable sources of impairments in assessed bays and estuaries.

| Ranking | Probable Source Category | Square Miles Threatened or Impaired |
|---|---|---|
| 1 | Atmospheric deposition | 8026 |
| 2 | Unknown | 5773 |
| 3 | Municipal discharges/sewage | 5021 |
| **4** | ***Industrial*** | **4239** |
| 5 | Other | 3825 |
| 6 | Natural/wildlife | 3521 |
| 7 | Spills/dumping | 3137 |
| 8 | Unspecified nonpoint source | 2833 |
| 9 | Agriculture | 2557 |
| 10 | Urban-related runoff/stormwater | 2249 |
| 11 | Habitat alterations (not directly related to hydromodification) | 2057 |
| 12 | Hydromodification | 1940 |
| 13 | Legacy/historical pollutants | 1516 |
| 14 | Resource extraction | 780 |
| 15 | Recreational boating & marinas | 525 |
| 16 | Commercial harbour & port activities | 495 |
| 17–22 | Sum of additional 7 minor categories, including aquaculture, recreation, military bases, construction etc. | 67 |

*Source:* USEPA accessed March 14, 2014.

A more efficient approach includes responding to regulation and economics by producing by-products rather than wastes (e.g., animal feed from spent grain from distilleries, biogas from sugar industries). That modern approach (D'Arcy *et al.* 1999) means that ***no longer*** will the cost of effluent treatment be determined by:

(pollutant load) × (cost of pollutant removal/$m^3$) = cost of effluent treatment operation (& size of treatment plant)

Instead a process of evaluation of use of resources, maximisation of primary product production and capture, resource optimisation and recovery, and potential for by-products (including energy and water) from what were formerly wastes, is the basis of a modern and more sustainable industrial business. Examples are given in Edwards and Johnston (1996), D'Arcy *et al.* (1999), and D'Arcy (1991). In order to facilitate the adoption of such a philosophy, it is essential that waste regulators do not require by-products to still be treated as wastes and subject to restrictive and bureaucratic waste regulations.

Serious problems remain however, especially in developing countries, where pollution history has often been repeated, with primitive attitudes to single-purpose production with all the inefficiencies that cause gross pollution. Examples include the heavily industrialised catchments of the Tiete and Cubatao rivers in Brazil, the chronic pollution problems of the Niger Delta Oilfields in Nigeria, and the environmental impacts of industrialisation in some parts of Asia (references in Tables 1.4–1.5). The economic consequences include destruction of water-dependent business opportunities for others downstream (e.g., food industries, local fisheries and tourism), as well as more obvious environmental and human impacts. Such characteristics, reminiscent of all the mistakes of early industrialisation two hundred years ago in Europe, may be compounded by the continuing pollution problems – still evident in Europe and the USA – associated with stormwater management, and with persistent pollutants.

Pollution of the water environment from industry is not simply a consequence of effluent process discharges or major accidents. The following sections briefly consider environmental impacts of industrial effluents, of diffuse pollution at point of manufacture/processing, and diffuse pollution at point of application of products and in use.

## 1.2 INDUSTRIAL EFFLUENT DISCHARGES

### 1.2.1 Industrial effluents

Detailed consideration of effluent treatment technology is outwith the scope of this book, since each industry has its own characteristics and hence specific detailed requirements. For effluents, this introductory paper only seeks to introduce the issues and some example pollution history, as a context for design considerations for modern industrial development.

An industrial or trade effluent is an aqueous waste stream, associated with an industrial process. The latter could be production of beer or spirits, industrial ethanol, bleaching textiles, cooling systems, washing plant, vehicles or premises, or refining crude oil to produce the spectrum of hydrocarbons for the petrochemicals industry, or any of many other activities which generate contaminated wastewater.

It is not unusual for water quality in rivers in industrialised countries to have been dominated by effluent discharges from industry, for example the estuaries of the Tees and the Mersey in England, UK (Porter, 1973), and other internationally notable examples in Table 1.4.

Sometimes a single industrial discharge alone was sufficient to severely degrade a river or estuary, for example the BOD load from a yeast factory discharged in the 1990s to the head of the estuary of the River Forth in Scotland. Industrial effluents accounted for 97% of the pollution, and one plant – a yeast factory at the head of the

**Table 1.4** Examples of pollution associated primarily with trade effluent discharges.

| Example | Pollutant | Activity | Causes | Comments | Example References |
|---|---|---|---|---|---|
| Minimata Disease, Japan, 1953+ | Organic Hg | Acetaldehyde production, Chisso Chemicals Plant | Effluent & bio-accumulation | Food-chain uptake: fish poisoned, people too; for decades | Harada, 1995 |
| Itai itai Disease, Japan, 1955 | Cadmium | Cadmium mining, Kamioka mine | Effluent | Mine-water polluted river used for rice irrigation | Yoshida *et al.* 1999 |
| Cuyahoga River, USA, 1952, 1969 | Oil | Oil refining, distribution & use | Effluent & accidents | River caught fire: led to environmental movement in USA; EPA /Ohio EPA | *Time*, 1969; Adler, 2002 |
| River Ou, China, 2014 | Oil | Oil refining, distribution & use | Effluent & accidents | River caught fire | *Daily Mail* Friday March 7th 2014, p. 34 |
| Mersey bird kill, UK, 1979–1982 | Alkyl lead | Manufactures of lead anti-knock petrol additive | Effluent & bio-accumulation | Food-chain uptake poisoning birds | Wilson *et al.* 1986 |
| Cubatao river pollution, Brazil, 1980s | Toxicity | Oil, steel, fertilisers & other industries | Effluents and incidents | Variously cited as one of the most polluted rivers | Zagatto *et al.* 1987; Wikipedia |
| Teite River, Brazil, 1980s–1990s | Many industrial pollutants | 40,000 industrial businesses | Effluents, disposal, spills | Sewage and industry; impacts 260 km d/s Sao Paulo | Alonso & Serpa, 1994; *Economist* 2011 |
| Loch Leven, UK, 1960s–1980s | Phosphate & Cl-H/C pesticide | Bleaching and producing woollens | Effluent | Accounted for almost a third of ortho-P load to lake | D'Arcy, 1991 |

estuary – dominated, accounting for downgrading 6.1 km$^2$ of the estuary to class 3, the second lowest quality ranking in a four-category classification scheme where 1 was excellent and 4 seriously polluted (FRPB, 1995). The Forth was restored to good quality, with the return of sensitive fish species such as sparling (*Osmerus eperlanus*) as clean up systems were introduced at the industrial premises (recovering value from waste, and treatment of residuals).

### 1.2.2 Mining industry

Mining is an interesting pollution source, since it is often intimately linked to the water environment. Panning for gold in-river (Tarras-Wahlberg *et al.* 2000; Bannister *et al.* 2006) has caused degradation of watercourses and surrounding landscapes on varying scales, especially when toxic chemicals (e.g., cyanide and mercury) are used to enhance recovery of the mineral (Pulles *et al.* 1996; Jones & Miller, 2004; San Francisco Estuary Institute [SFEI], 2015). Surface mining exposes earth to wind and rain erosion and exhibits therefore the characteristics of diffuse pollution, including risks of oil spills or leaks from machinery. Deep mines can result in a legacy of pollution long after the mine closes, due to chemical changes in the strata during the period when groundwater was being continuously pumped from the mine to allow extraction of the resource, whether gold, coal or other minerals (Younger. 2000; Dowson, 2003; Heath *et al.* 2009). Even quarrying silica sand can cause low pH drainage; arising from sub-soil overburden dumps, when sulphides are oxidised and weak sulphuric acid seeps from the dump in wet weather, unbuffered by calcium minerals. Examples of pollution associated with the mining industry are given in Table 1.5.

In Scotland there were some 560 coal mines in the 1800s; a hundred years later there were 218, of which 110 were in the Forth catchment. By 1995 only 2 remained. Yet 22% of the downgraded reaches in the freshwater river Forth, was accounted for by mining (FRPB, 1995). The great majority of those impacts by then were due to water table rebound, when groundwater re-filled the abandoned mine workings, leaching ferrous sulphate, produced during the active mining period of dry conditions, when naturally present insoluble sulphide in the rocks became oxidised to (soluble) ferrous sulphate. On mixing with river water as the groundwater emerged as a spring (often not at the former effluent discharge point), the ferrous sulphate removed oxygen from the water and precipitated as orange ochre – insoluble ferric sulphate which characterises the ferruginous discharges across the UK in former coal mining catchments (Younger, 2000). Trade effluent discharges ceased, and diffuse pollution increased.

### 1.2.3 Effluent impacts case study: The River Mersey Bird Mortality

The River Mersey is formed from the confluence of industrialised tributary rivers which arise in the Pennine hills above Manchester, in NW England. The river runs

**Table 1.5** Pollution examples from the mining industry.

| Example | Pollutant | Activity | Causes | Comments | Example References |
|---|---|---|---|---|---|
| Wheal Jane, Cornwall, UK 1992 | Iron & other trace metals, low pH | Abandoned tin mine | Rising groundwater after mine closure | Carnon River & Fal Estuary discoloured; peak contamination of metals 900xEQS Zn | Younger *et al.* 2005 |
| King River, Tasmania | Copper, sulphidic acids | Copper mining | Effluent | | Locher, 1995; Dowson, 2003 |
| Phalaborwa Mine, South Africa | Low pH, metals, aquatic toxicity | Copper, zinc & phosphate mining | Dust, tailings dams overflowing, discharges | Rivers in the Kruger National Park (KNP), bioaccumulation in aquatic organisms, water not drinkable at camps in KNP | Heath & Claassen, 1999; Kotze *et al.* 1999 |
| Blesbokspruit South Africa (gold mining, & paper factory) | Sulphate (salts), iron and trace metals | Abandoned mine & closure of pulp & paper factory | Rising groundwater after mining ceased & no appropriate closure or responsibility for liability | Acid mine drainage required treatment to protect a Ramsar wetland | Ambani & Annegarn, 2015 |
| Ecuador 2000 | Mercury & cyanide, trace metals | Artisan gold mining | Primitive mining and processing | Extensive impacts, small scale | Tarras-Wahlberg *et al.* 2000 |
| Mining impacts, UK | Iron | Mainly abandoned coal mines | Rising groundwater after mine closure | 400–600 km affected by ferruginous drainage, UK | Younger, 2000 |

from Manchester for 113 km to the sea in Liverpool Bay. The Mersey Estuary has an extensive inner basin, flanked on the south side by the Manchester Ship Canal, which received effluent from a lead anti-knock (tetra-alkyl lead) plant in Ellesmere Port (Figure 1.1). Following a steadily improving trend in water quality during the 1980s, fish began to return to the estuary (Wilson *et al.* 1986) and the estuary supported large numbers of over-wintering wildfowl and wading birds.

In September 1979, sick and dead wading birds, plus a handful of waterfowl, began to be noticed by bird watchers at a ringing station in Hale, on the north bank of the upper estuary, and by wildfowlers on the marshes along the south bank. Both local groups stated that the phenomenon was highly unusual and they suspected pollution. Analysis of the dead birds revealed elevated concentrations of lead in liver and other tissues. But the concentrations were lower than in some estuaries draining lead-mining catchments (e.g., Gannel in Cornwall), where there were no bird mortalities.

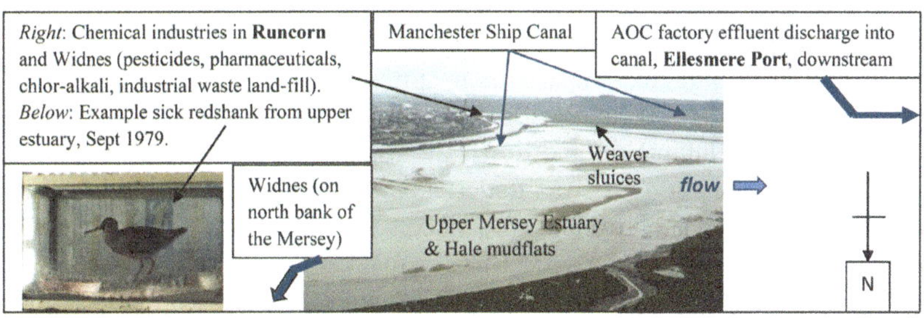


**Figure 1.1** Photo-diagram illustrating the scene of the 1979 Mersey Bird Mortality, NW England, UK (Photo credits: BJ D'Arcy).

In any such investigation, it is by no means always clear which are the chemicals to be investigated in an effluent brew of product, process side reactants, raw materials and the spectrum of forms in between. Associated Octel Company (AOC), the manufacturer of the lead anti-knock product, conducted their own investigations and analysis for their product and its tri-alkyl lead water soluble, stable, break-down derivatives. They informed the investigators in North West Water of their belief that the pollutant affecting the birds was tri-alkyl lead. The bird mortalities continued through October 1979, ceasing in November. The company co-operated fully with the investigations and no-one was initially able to explain why the incident should occur suddenly in 1979. Subsequent laboratory tests dosing starlings with alkyl lead replicated the symptoms observed in the dying birds, and post-mortem analysis of the starlings revealed comparable tissue concentrations of lead to those found in the dead/sick wading birds from the Mersey (Osborne *et al.* 1983). The organic form of lead was not only water soluble, but more fat soluble than inorganic lead. That greater availability seemed likely to explain why mortalities occurred at elevated total lead concentrations, but lower than might be expected from estuaries polluted by inorganic lead from historic mining activity. Why did this occur in 1979? One theory was that the spat 'bloom' of the small estuarine shellfish *Macoma balthica* which was noted in the upper estuary, provided a preferred food species which bioaccumulated the lead, replacing the more usually available (in the Mersey) worms in the diet of the birds; the shellfish had five times the lead concentration of the worms (Maddock & Taylor, 1980; Wilson *et al.* 1986). The last detail which needed to be resolved was the fact that the AOC discharge was into the tidal reach of the Manchester Ship Canal, (MSC), not the Mersey. The environmental compartments from process effluent to biota were:

Process effluent → MSC water → Estuary water → biota → birds
→ man (wildfowlers)

Earlier investigations had shown that the Ship Canal only passes relatively small flows out into the Estuary during the drier summer months (and neap tides), via lock gates at the mouth of the canal. But when the largest tides (spring and autumn) equal or exceed the design level in the canal, lock gates at the canal mouth and also upstream in Runcorn are opened and draw water – with the accumulated daily load of alkyl lead from AOC – *upstream* to spill into the upper estuary at Runcorn (D'Arcy & Wilson, 1978). That process is in addition to the normal tidal movements and dispersal of pollutants up and down the estuary on every tide. A review of the discharge permit for AOC was undertaken and new limits agreed, based on correlations between dead birds and effluent concentrations, plus a safety allowance for possible sub-lethal effects on birds. Birds were taken as the critical environmental factor (see Table 1.6); the evidence predicted an increasing risk of bird mortalities at a discharge in excess of 6 mg/l soluble lead. Similar levels would be likely to result in liver concentrations in excess of the 0.5 mg/kg threshold identified by Osborne *et al.* (1983). The limiting load was therefore agreed with the discharger, and target environmental quality standards for soluble lead were derived for estuary water.

**Table 1.6** Correlation coefficients for alternative regressions ($y = mx + c$) for levels of alkyl lead in various environment compartments.

| Y | X | R | n |
|---|---|---|---|
| No. dead birds (all spp.) | Concentration in *Macoma balthica* | 0.90 | 14 |
| Concentration in Mersey Estuary water | Discharge load | 0.87 | 48 |
| No. dead birds (all spp.) | Discharge concentration (mean previous month) | 0.81 | 17 |
| No. dead birds (excluding gulls) | Discharge concentration (mean previous month) | 0.77 | 17 |
| Concentration in MSC | Discharge concentration | 0.76 | 27 |
| No. dead birds (all spp.) | Discharge concentration (monthly mean) | 0.71 | 17 |
| Concentration in *Macoma balthica* | Estuary water concentration | 0.68 | 288 |
| Concentration in *Macoma balthica* | Discharge load | 0.65 | 288 |

*Source:* Adapted from Wilson *et al.* (1986).

When water quality standards were exceeded in subsequent years, as predicted dead and sick birds again began to be found. This case study illustrates the difficulties in protecting the water environment when regulating a large number of potentially toxic effluents discharged to a large complex water body and investigating and preventing pollution problems, even with the full co-operation of the industries there. It should be noted that alternative theories were put forward to explain the mortality (Smith *et al.* 1982), but were not given priority over the case for tightening controls at the AOC factory (by the company and the regulator). Process controls improved discharge quality consistently by 1983 (D'Arcy *et al.* 1999). The plant is still operational, reprocessing tank farm sludge contaminated with alkyl lead, from petrochemical plants around the world (AW Wither, pers. com. 2015).

## 1.3 INDUSTRIAL COMPLEXES – POINT SOURCE AND DIFFUSE POLLUTION

### 1.3.1 Drainage from industrial districts

Industrial cities often present serious problems for water quality and resources (Banerjee, 2015). In any large industrial district, there will probably be some trade effluent discharges (process effluents, aqueous waste streams, also known as point sources) hopefully with wastewater treatment plants (WWTPs), or agreed connections to a foul sewer for treatment in admixture with sewage. In addition, the surface runoff in drainage from the area will be contaminated by multiple lesser sources of pollutants, mobilised by rainfall to discharge to a stormwater sewer or watercourse, and/or seep into groundwater: diffuse pollution. There is often historical contamination too: in the UK and elsewhere, as older industries declined and closed, their process effluents ceased too. But often contaminated land remained, contributing polluted rainfall runoff, and seepage to groundwater contaminating resources for many years after closure. Oil and chemical industries, but also railway marshalling yards, timber treatment sites, ship-yards using anti-fouling paint, engineering factories and power generation installations using transformer oil, have all been implicated in either groundwater or surface water pollution by toxic and sometimes persistent pollutants. Consequently, estimating industrial loads, for example toxic metals, is uncertain (Behrendt *et al.* 2005).

In South Korea, data has been reviewed to allow a comparison of quality in industrial discharges, urban runoff, and industrial streams (diffuse sources). Table 1.7 compares average pollutant concentrations measured in effluent from industrial WWTP discharges, with average concentrations measured in industrial streams and urban streams, and in rivers and lakes. The industrial streams do not include the industrial WWTP discharges. The average levels of contamination were highest from the industrial wastewater treatment plants, with urban streams next highest or the same as the industrial streams. In the wet season, concentrations of the major point source effluent discharges reported in Table 1.7 decreased, due to dilution from other areas. But *loads* of diffuse source pollutants measured in the industrial estate streams actually *increased* with wet weather, mobilised by rainfall.

**Table 1.7** Average pollutant concentrations for effluents and water environments in S. Korea in 2012.

| Parameters | Industrial WWTP Discharges | Industrial Streams | Urban Streams | Rivers | Lakes |
|---|---|---|---|---|---|
| No. of monitoring stations | 69 | 70 | 49 | 840 | 189 |
| pH | 7.44 | 7.70 | 7.68 | 7.86 | 7.83 |
| BOD (mg/l) | 34.87 | 8.80 | 12.18 | 2.14 | 1.79 |
| COD (mg/l) | 28.32 | 11.00 | 10.84 | 4.91 | 4.40 |
| TOC (mg/l) | | | 6.17 | 3.13 | 2.68 |
| SS (mg/l) | 19.58 | 13.40 | 13.48 | 11.02 | 7.81 |
| Temp. (°C) | 19.70 | | 16.51 | 14.71 | 14.07 |
| EC (μmho/cm) | 3315 | 6612 | 4623 | 535 | 1581 |
| TN (mg/l) | 14.88 | 8.30 | 8.50 | 3.58 | 1.97 |
| NH$^{3}$N (mg/l) | | | 3.72 | 0.52 | 0.15 |
| DTN (mg/l) | | | 7.83 | 3.37 | 1.79 |
| T-P (mg/l) | 0.84 | 0.29 | 0.51 | 0.10 | 0.04 |
| DTP (mg/l) | | | 0.53 | 0.09 | 0.03 |
| Chl. A (mg/l) | | | 11.95 | 12.26 | 13.68 |
| Cd (μg/l) | 112 | 1.30 | | ND | ND |
| CN (μg/l) | 11.69 | 7.20 | | ND | ND |
| Pb (μg/l) | 2.46 | 6.00 | | ND | ND |
| Cr+6 (μg/l) | 0.35 | 0.80 | | ND | ND |
| Cu (μg/l) | 62 | 24 | | ND | ND |
| Zn (μg/l) | 166 | 273 | | ND | ND |
| As (μg/l) | 2.77 | 2.10 | | ND | ND |
| ABS (μg/l) | 221 | 105 | | ND | ND |
| Phenol (μg/l) | 5.97 | 1.00 | 0.15 | ND | ND |
| Nor. Hexane (μg/l) | 1138 | 271 | | ND | ND |
| Faecal Coliform (MPN/100ml) | | | 672219.59 | 6624.22 | 390.98 |
| Total Coliform (MPN/100ml) | 108,179 | 797,890 | 3,991,108 | 27,683 | 3273 |

## 1.3.2 Major accidents and other industrial incidents

Most industry is now located in business parks and industrial estates (see specific paper; 1.3), with relatively few industries having direct discharges of process effluents to the water environment. Even 'clean' industries however, can be serious polluters if accidents involving raw materials, intermediates or final products occur, or in the event of fires and loss of pollutants entrained in firewater. Many of these estates are very large, with many companies of all kinds spread over extensive areas. All have surface water drainage and stormwater runoff is often contaminated by many diffuse sources. Pollution of the water environment from such extensive industrial and commercial zones is diffuse in nature, which is not to say that there are not hotspots where pollutants occur in greater quantities, and some activities present far more risk than others. Some are *ad hoc* trade effluents, such as mobile steam cleaning units, which are very difficult to regulate. There is a complete spectrum of pollution incidents and accidents from the individually trivial but often regular (e.g., drips of oil, or chemicals gradually accumulating on the ground and in drains) to large scale intermittent spills, or inadequate disposal of potentially polluting wastes. Fires are another major pollution risk category, when firewater mobilises stored pollutants and contaminants arising from combustion of the premises. Flooding may also move pollutants into the water environment and is another important risk.

The examples in Table 1.8 include pesticides lost from shipping incidents, which receive less media attention than the better-known oil tanker oil spills or off-shore incidents such as the Gulf of Mexico.

There is now a great deal of experience and technical guidance to minimise risks of major pollution incidents. Even in the developed countries however, sometimes business dogma incurs unnecessary expense in compensation for avoidable pollution, despite technical knowledge to easily prevent it (e.g., groundwater pollution at Hinkley, USA, perhaps the BP oil leak in the Gulf of Mexico too, where 'savings' resulted in massive extra costs and loss of confidence in the company (*Economist*, 2015, 2010)).

## 1.4 DIFFUSE SOURCES OF POLLUTION

### 1.4.1 Diffuse pollution at point of manufacture or processing

Major and minor incidents are in addition to anthropogenic background contamination arising from combustion of fuel, power generation, use of commercial vehicles and machinery. This includes roof runoff contaminated by extractor fans in machine workshops discharging very fine particulate pollutants onto roof areas. Diffuse pollution hotspots occur when polluting activities result in significant contamination at particular premises, or areas of ground or roofs thereon, and pollutants are liable to be mobilised by rainfall and washed into drainage networks or otherwise into the aquatic environment. In a study in India (Joshi & Obaidy, 2014), comparison of runoff-rainfall quality ratios found highest values from industrial land-uses, for organic pollution, with the next highest being from residential areas. Studies in Europe and North America have also often shown industrial areas to be characterised by higher concentrations of a greater variety of pollutants in urban runoff, although traffic related pollutants on

**Table 1.8** Examples of leaks, accidents, and fires as ways that industrial activity can pollute the water environment (major incidents).

| Example | Pollutant | Activity | Causes | Comments | Example References |
|---|---|---|---|---|---|
| Niger delta, Nigeria | Oil | Oil extraction and refining | Leaks & incidents, pipeline explosion | Fish mortalities, loss of clean water resources (potable supplies, fisheries, health) | Nwankwo & Irrechukwu, 1987; Omodanisi *et al.* 2014 |
| River Rhine, Switzerland 1986 | Pesticides, mercury | Pesticides production and storage | Fire: loss of pollutants in fire-water runoff | Fish mortalities, sediment dredging to remove toxic chemicals | Salzmann, 1987; Van Urk & Kerkum, 1987; Giger, 2009 |
| Hinkley, California 1952–66 | Hexavalent chromium | Storage of contaminated water in un-sealed pits | Inadequate storage | Groundwater resource polluted | California EPA, 2016 |
| River Dee, Wales, UK | Phenol | Storage vessel on industrial estate | Leak | Water supply for 2 million people cut-off for 2 days. Incidents at same estate remain a risk | D'Arcy *et al.* 2000 |
| Birkenhead docks, River Mersey | Mollases | Food industry tank farm | Burst pipe | Dissolved oxygen in the dock dropped rapidly over days | Unpublished report, North West Water |
| River Bourne, Trib. Thames, UK, 1990 | Lindane & tributyl tin oxide | Timber treatment | Leak from storage tank | 15,000 fish killed, water supply intakes closed for days | Surveyor, 1990; Dowson *et al.* 1996 |
| Shipping | 6 tonnes of lindane, 1.6 t permethrins | Transport of pesticides as deck cargo | Loss of *MV Prentis* in English Channel | Pesticides in containers on deck. | Johnston *et al.* 1997 |
| Samarco, Brazil, Nov. 2015 | Ferrginous drainage, mud & trace metals | Iron ore opencast mine | Tailings dam burst | 12–15 deaths, as well as environmental devastation | Mining.com http://www.mining.com/several-casualties-after-tailings-dam-bursts-in-brazil/ |

major highways can exceed some industrial/commercial values. Many of the pollutants of most concern on industrial estates however, are not subject to routine water quality monitoring and hence data on impacts can overlook the importance of these pollutant sources, (e.g., example solvents, detergents, hydrocarbons and pesticides, as well as toxic metals). As well as adversely affecting surface waters, these pollutants have been identified as serious risks to groundwater quality, with long-term impacts on water resources. Some examples are given in Table 1.9.

**Table 1.9** Examples of local diffuse pollution impacts from industrial estates and premises.

| Example | Pollutant | Product | Pathway | Comments | References |
|---|---|---|---|---|---|
| Industrial estate on R. Dee, N Wales, UK | Phenol & other chemicals | Various | Surface water runoff | >2M potable water consumers at risk | D'Arcy *et al.* 2000 |
| Industrial estate, UK | Chlorinated hydrocarbons | Solvents | Fractured chalk strata | Diffuse sources | Lawrence *et al.* 1996 |
| Industrial estates in NW England, 1980s | Many, e.g., oil, toxic metals, BOD | Various; food, engineering, manufacturing | Surface water sewers, 22 industrial estates | Class deteriorations in water quality d/s industrial estates | D'Arcy *et al.* 1985 |
| Roorkee town, Uttarakhand, India | BOD, toxic metals, & others | Various industrial & commercial businesses | Surface water runoff | Highest BOD from industrial areas, & for heavy metals also commercial & transportation | Joshi & Obaidy, 2014 |
| Caw Burn, Scotland, UK | Oil, BOD, toxic metals | Various from large industrial estate, Livingston | Surface water runoff to separate sewers | Retrofit small 'first flush' wetland achieved class improvement | Heal *et al.* 2005 |
| Airports, UK | | De-icing chemicals | | | D'Arcy *et al.* 2000 |

## 1.4.2 Industrial products causing pollution at point of use

There is another, more widespread form of diffuse source pollution involving industrial chemicals: pollution associated with the points of use or consumption, often remote from the places of manufacture (Table 1.10). For example, brake pads for vehicles may be manufactured at an industrial plant on an estate in a city, but

**Table 1.10** Examples of industrial pollution associated with use of products (arising at point of consumption not production; often widely dispersed, impact may be distant from point of use as well as manufacture).

| Example | Pollutant | Product | Pathway | Comments | References |
|---|---|---|---|---|---|
| Great Lakes, North America | POPs, PCBs | Hydrocarbon fuels (on combustion) | Air pollution & precipitation | Contamination of water environment | Hornbuckle & Green, 2003 |
| Military herbicides | Dioxin | Agent Orange | Bioaccumulation: plants, soil, pond sediments, fish, ducks, people | Defoliant used in Vietnam | Dwernychuk, 2002 |
| Imposex | Oestrogens & analogues, some pesticides alkyl tin | Anti-fouling paints, and other hormone analogues | Direct application to hulls of boats and ships | Also other compounds implicated, e.g., some pesticides. | Harrison, 2001 |
| Top of food-chain fauna, e.g., San Francisco Estuary | PCBs | Transformer oil, road sealant, & other uses | Leaks and combustion losses to atmosphere | Urban stormwater largest pathway | SFEI, 2015 |
| Fin whales, Mediterranean 2003 | PCBs, PAHs, DDT | Various | Runoff & atmos. ppt. | Shipping and other incidents too | Notarbartolo-di-Sciara *et al.* 2003 |

| | | | | | |
|---|---|---|---|---|---|
| Otters, *L. lutra* | Organochlorine pesticides | Probably several products | Bioaccumulation in food chain from application to land, then in runoff | Recovery eventually, where most uses banned (e.g., UK) | |
| Polar bears | DDT, Dieldrin, POPs, PCBs, chlorinated hydrocarbons | Persistent pesticides, fire retardants, & others | Food chain Concentration | Global distillation | Kucklick *et al.* 2002; Pedersen *et al.* 2015 |
| Toxicity in estuarine/ marine environment | Copper | Brake pads, Roofing materials, etc. | Wear, dust deposition, wash-off | Brake pad initiative & other actions | SFEI, 2015 |
| Road traffic | Zn, Cu, and other metals, PAHs, Oil | Vehicle components, road surfaces & associated structures | Wear, dust deposition, wash-off | | Napier *et al.* 2008 |
| Ingestion of plastic by fur seals | Plastic particles & fragments, microplastics, nanoparticles | Miscellaneous use and disposal including in cosmetics | Contamination of seas and oceans; partial break-down | 164 plastic particles in 145 seal scats; Macquarie Island | Eriksson & Burton, 2003 |

the pollution derives from the wear on each pad on each vehicle driven in the catchment, finely divided dust from wear settling onto roads and other surfaces, washing into drainage networks and hence the water environment. Similarly, engine wear for other toxic metals, tyres for metals plus polyaromatic hydrocarbons (PAHs) and other matter (Napier *et al.* 2008). Air emissions as well as transport in rivers and streams can carry pollutants far, as can winds and ocean currents, and bioaccumulation in places remote from sources may become an issue for predators (including humans) at the top of food chains (e.g., birds of prey, otters, seals and polar bears). The biota may be migratory too.

Two categories of diffuse pollution causes are evident in Table 1.9:

(1) The deliberate creation by industry of a toxic material designed to be introduced to the environment to kill or suppress living organisms.
(2) The production of industrial chemicals for uses that do not require any environmental toxicity and for which environmental impacts when in widespread use had not been foreseen.

Direct application introduced serious toxic pollutants to the environment, for example pesticides in agriculture, forestry and urban environments, anti-fouling paints to reduce fuel consumption for ships, and military uses (for example Agent Orange defoliant used in Vietnam, Laos and Cambodia).

The second category includes PCBs in transformer oil, lead anti-knock in petrol additives, solvents, detergents, fire-retardants, and copper in brake pads. Toxicity was not a sought-after property for those purposes, yet they have caused widespread environmental contamination. The global use of plastic has now contaminated seas and oceans everywhere; break-down products of small fragments of plastic have been found in the droppings of fur seals in the southern oceans (Eriksson & Burton, 2003). For this category, diffuse pollution can be seen as the environmental science of unintended consequences. 'Green Chemistry' aims to address that challenge and is: 'the design of chemical products and processes that reduce or eliminate the use or generation of hazardous substances' (USEPA, 2017). The challenge is great, but positive engagement with industry offers more prospects for proactive action by industry than efforts simply to regulate once a problem has become indisputable (e.g., www.epa.gov/greenchemistry/presidential-green-chemistry-challenge-award-recipients-industry-sector).

## 1.5 DISCUSSION

Industrial impacts are still severe and destructive at many places of manufacture, in many parts of the world. Yet technology is well developed now for far more efficient industrial processes and optimising value from materials. Is the problem simply business dogma at a political level, a high level (remote from production experience in modern industries) prevalence of the belief that environment is the enemy of business profit? Or is it linked to business models seeking quick returns, without investment in additional plant and processing? (or in health, safety and environment)? Some technologies for maximising returns per unit of throughput do have a longer pay-back by comparison with quick 'extract a small amount and sell it with the rest to waste' approaches. Adler (2002) in his interesting account of the oil industry in Cleveland notes that before there was a market for gasoline, the quantity of waste oil from which it was later a derivative, was simply dumped in the river. There is global expertise available at least in theory, with the increasing globalisation of industries. But to date there seems a mixed picture and inadequate evidence of export of modern efficient and safe production processes. A more sustainable business can maximise value per unit of resource consumed and sustain a healthy workforce and local environment, contributing to the economy of the country in which it is located. Perhaps there is a greater role for trade tariffs, for example by powerful economic groups such as the European Union, discriminating against polluting industries, on a free trade for non-polluters basis?

National governments and regional or city authorities have an important role to play (again see Adler, 2002). Local industrial impacts can severely constrain other economic activity in a region. But that is an externality for the business concerned, since the destruction of wealth generation for others is arguably not the concern of the polluters. Enforcement of 'the polluter pays' rationale, developed strongly in the EU for example, is a matter for government regulation and especially enforcement.

The effects of industrial products as contaminants is perhaps technically more difficult. Until product substitution can achieve replacement of many toxic materials with biodegradable or inert and safe alternatives, controls on secondary (associated with use and disposal) emissions must remain a principal measure for environmental protection. Measures to alter the composition of plastic bags have improved but are still inadequate. More progress has been made in recently with initiatives to reduce demand (e.g., tax on disposable plastic shopping bags in UK).

Some pollutants can never be biodegradable of course. New industries for essential but toxic metals for example may perhaps be developed for smaller scale recovery of resources from urban waste streams (higher concentrations per kg of municipal waste than in many ores), with economics linked to waste disposal business as well as inherent value of recovered metals (at scales which match demand).

There remains a fundamental dilemma for control of many pests – chemicals applied to the landscape to control a pest species for economic or human health benefit may destroy a far wider array of species and even risk the health of local people and livestock. The succession of classes of pesticides from organo-chlorine to organo-phosphates to permethrins is a journey which will continue, but perhaps more sophisticated biological techniques to control pests or vectors by ways which are far more species-specific, will be the future? Similarly, how can we protect the longevity of clothes, fabrics and timber products, whilst ensuring that when worn out or demolished they can be recycled or degraded? Mothproofers and timber treatment chemicals are some of the most polluting materials known.

It is one thing to control a disease or protect a food crop, quite another to apply toxic materials to control a personal preference for an aesthetic appearance. For example, there have been moves to add copper to roofing felt to prevent moss growth, or the use of copper sheeting on buildings for aesthetic appearance. Copper is a relatively water soluble toxic metal, especially where exposed to unbuffered rainfall. Can households, building suppliers and architects be encouraged to make more environmentally safe choices?

Locally diffuse source pollutants can be captured in drainage networks, to protect the receiving water environment. For degradable and ubiquitous pollutants such as oil, it is sensible to encourage the widespread use of green infrastructure measures, variously known as urban best management practices (BMPs), sustainable urban drainage systems (SUDS) or low impact development (LID) techniques.

Local controls on uses, for example traffic reduction measures, housing programmes, regulation of waste burning, will not just protect air quality locally, and reduce the contaminant burden on green infrastructure, but will also help reduce global contamination associated with air movements and precipitation at points distant from sources.

## 1.6 CONCLUSIONS

Industry has been and remains a significant source of environmental pollution. Pollution of the water environment involves four main categories of sources:

(i) Direct discharges of trade (process) effluents into local water bodies
(ii) Precipitation of pollutants arising from atmospheric emissions
(iii) Diffuse pollution at manufacturing and handling/storage sites; may involve catastrophic accidents and fires, or simply cumulatively important chronic pollution from contaminated runoff
(iv) Diffuse pollution from products when in use, often far from source of production.

Pathways may be direct to watercourse, indirect via stormwater drainage networks, or seepage into groundwater from contaminated land, or by atmospheric/land interactive processes such as global distillation.

Inefficiency, and lack of commercial maximisation of value from raw materials, stored products and wastes, has been a common characteristic where industrial pollution is evident at factories and industrial estates; almost literally ‘money down the drain’.

Impacts may be rare but catastrophic (e.g., large fires and accidents) on any water body, from clean rivers to urban/industrial watercourses, or intermittent and multiple sources resulting in chronic poor water quality. Capture and, where possible, degradation in green infrastructure or other SUDS features offers a practical solution for existing diffuse contamination, after housekeeping measures at high risk areas.

Some diffuse pollution by industrial products is foreseeable and hence controllable, for example the successive introduction and withdrawal from use of classes of pesticides, and more targeted applications. Persistent and toxic pollutants associated with other industrial products are more difficult to anticipate and control, and have resulted in global contamination, (e.g., PCBs, fire-proofing chemicals), with hotspots in top predators as a result of concentration in food chains. The current focus on priority and emerging pollutants is a necessary response to this challenge.

## 1.7 ACKNOWLEDGEMENTS

Thanks to Tom Leatherland, Gordon Mitchell, Ralph Heath and Jean-luc Bertrand-Krajewski for useful comments on the paper.

## REFERENCES

Adler J. H. (2002). Fables of the Cuyahoga: reconstructing a history of environmental protection. *Fordham Environmental Law Journal*, **89**. Faculty Publications, paper 191, http://scholarlycommons.law.case.edu/faculty_publications/191.

Alonso L. R. and Serpa E. L. (1994). O Controle da Poluição Industrial no Projeto Tietê, 1994. Companhia de Tecnologia de Saneamento Ambiental (CETESB), São Paulo.

Ambani A.-E. and Annegarn H. (2015). A reduction in mining and industrial effluents in the Blesbokspruit Ramsar wetland, South Africa: has the quality of the surface water in the wetland improved? *Water SA*, **41**(5) Pretoria Oct. 2015. http://dx.doi.org/10.4314/WSA.V41I5.08. On-line version ISSN 1816-7950, Print version ISSN 0378-4738.

Banerjee S., Maiti S. K. and Kumar A. (2015). Metal contamination in water and bioaccumulation of metals in the planktons, molluscs and fishes in Jamshedpur stretch of Subarnarekha River of Chotanagpur plateau, India.

Bannister S., Heath R. G. M. and Pulles W. (2006). Small scale mining Best Practice Guideline. Report, Department of Water Affairs and Forestry, 2006, Best Practice Guideline: A1.1, South Africa.

Behrendt H., Korol R., Opitz D. and Stronska M. (2005). Present state of heavy metal inputs in the Odra basin. In: Nutrients and Heavy Metals in the Odra River System, H. Behrendt and R. Dannowski (eds), Weissensee Verlag, Berlin, Germany, pp. 232–246. ISBN 3-89998-046-8.

California EPA (2016). http://www.swrcb.ca.gov/rwqcb6/water_issues/projects/pge/index.shtml; http://www.swrcb.ca.gov/rwqcb6/water_issues/projects/pge/index.shtml and http://geotracker.waterboards.ca.gov/.

D'Arcy B. J. (1988). How bad is the Mersey? *Environment Now*, Holmes McDougall Ltd., Glasgow, UK, No. 10, November, pp. 22–24. ISSN 0952-2352.

D'Arcy B. J. (1991). Legislation and the control of dyehouse pollution. *Journal of the Society of Dyers and Colourists*, **107**, 387–389.

D'Arcy B. J. and Wilson K. W. (1978). Fish surveys on the tidal Manchester Ship Canal – the implications for routine chemical sampling. Tech. Rept. TS/BS/78-2. North West Water Authority, Warrington, 26p.

D'Arcy B. J., Ellis J. B., Chatfield P., Andrew D. and Usman F. (2000). Chapt. 12: Chemicals. In: Diffuse Pollution Impacts, J. B. D'Arcy, J. B. Ellis, R. C. Ferrier, A. Jenkins and R. Dils (eds), CIWEM, London, pp. 113–122, ISBN: 1 870752 46 5.

D'Arcy B. J., Todd R. B. and Wither A. W. (1999). Industrial effluent control and waste minimisation: case studies by UK regulators. *Water Science and Technology*, **39**(10), 281–287.

Dowson G. (2003). King River case study update in State of the Environment Tasmania 2003. Resource Planning and Development Commission. http://www.rpdc.tas.gov.au/soer (accessed 14 February 2014).

Dowson P., Scrimshaw M. S., Nasir J. M., Bubb J. N. and Lester J. N. (1996). The environmental impact of a chemical spill from a timber-treatment works on a lowland river system. *Journal of Chartered Institution of Water and Environmental Management*, **10**, 235–244.

Dwernychuk L. W., Cau H. D., Hatfield C. T., Boivin T. G., Hung T. M., Dung P. T. and Thai N. D. (2002). Dioxin reservoirs in southern Vietnam – a legacy of Agent Orange. *Chemosphere*, **47**, 117–137.

Economist (2010) BP counts the political and financial cost of Deepwater Horizon Jun 17th 2010. From the print edition http://www.economist.com/node/16381032.

Economist (2010) The silvery Tietê. Cleaning up an open sewer Oct 22nd 2011, São Paulo. From the print edition http://www.economist.com/node/21533415.

Economist (2015). The final bill for the tragedy is now clear, but the costs of American litigiousness will be felt for years to come, Jul 2nd 2015, 16:56. *Business and Finance*. http://www.economist.com/news/business-and-finance/21656847-costly-mistake?zid=313&ah=fe2aac0b11adef572d67aed9273b6e55.

Edwards and Johnston (1996). Water and wastewater minimization: the Aire and Calder Project. *Journal of Chartered Institution of Water and Environmental Management*, **10**, 227–234.

Eriksson C. and Burton H. (2003). Origins and biological accumulation of small plastic particles in fur seals from Macquarie Island. *Ambio*, **32**(6), 380–384.

FRPB (1995). *Water Quality Objectives for 2000 and 2015*. Forth River Purification Board, Edinburgh.

Giger W. (2009). The Rhine red, the fish dead – the 1986 Schweizerhalle disaster, a retrospect and long-term assessment. *Environment Science and Pollution Research*, **16**(Suppl 1), S98–S111.

Harada M. (1995). Minamata disease: methylmercury poisoning in Japan caused by environmental pollution. *Critical Reviews in Toxicology*, **25**(1), 1–24.

Harrison P. T. C. H. (2001). Health effects of environmental chemicals. In: Pollution Causes, Effects and Control, R. M. Harrison (ed.), 4th edn, Royal Society of Chemistry, Cambridge, UK, pp. 500–523.

Heal K. V., Scholz M., Wilby N. and Homer B. (2005). The Caw Burn SUDS: Performance of a Settlement Pond/Wetland SUDS Retrofit. Proceedings of the Third National Conference on Sustainable Drainage, Coventry University, 20–21 June 2005, ISBN: 1 84600 007 6.

Heath R. G. M. and Claassen M. C. (1999). An overview of pesticides and heavy metal levels present in populations of the larger indigenous fish species in selected southern African rivers, WRC Report No. 418/1/99, 318pp.

Heath R. G., van Zyl H. D., Schutte C. F. and Schoeman J. J. (2009). First-order assessment of the quantity and quality of non-point sources of pollution associated with industrial, mining and power generation, WRC report No. TT 452/10, South Africa.

Hornbuckle K. C. and Green M. L. (2003). The impact of an urban-industrial region on the magnitude and variability of persistent organic pollutant deposition to Lake Michigan. *Ambio*, **32**(6), 406–411.

Johnston P., Marquardt S., Keys J. and Jeweel T. (1997). Shipping and handling of pesticide cargoes: the need for change. *Journal of Chartered Institution of Water and Environmental Management*, **11**, 157–163.

Jones G. and Miller G. (2004). Department of Natural Resources and Environmental Sciences, Final Report to U.S. Environmental Protection Agency Region IX Mercury and Modern Gold Mining, Mail Stop 199 University of Nevada Reno, NV 89557.

Joshi H. and Al Obaidy A. H. M. J. (2014). Adverse impacts of unplanned urban growth on the environment. *Water Science & Technology*, **70**(11), 1721–1728.

Kay J. P. (1832). The moral and physical condition of the Working Classes … in Manchester, pp. 12–34. In: Human Documents of the Industrial Revolution in Britain, E. R. Pike (ed.), George Allen & Unwin Ltd., London, 1966, pp. 309–312.

Kotze P. J., du Preez H. H. and van Vuren J. H. J. (1999). Bioaccumulation of copper and zinc in *Oreochromis mossambicus* and *Clarius gariepinus* from the Olifants River, Mpumulanga South Africa. *Water SA*, **25**(1), 99–110.

Kucklick J. R., Struntz W. D. J., Becker P. R., York G. W., O'Hara T. M. and Bohonowych J. E. (2002). Persistent organochlorine pollutants in ringed seals and polar bears collected from northern Alaska. *The Science of the Total Environment*, **287**, 45–59.

Lawrence A. R., Stuart M. E., Barker J. A. and Tester D. J. (1996). Contamination of chalk groundwater by chlorinated solvents: a case study of deep penetration by non-aqueous phase liquids. *Journal of Chartered Institution of Water and Environmental Management*, **10**, 263–272.

Locher H. (1995). Sediment Transport in the King River, Tasmania, Working Document 95/5. Cooperative Research Centre for Catchment Hydrology, Melbourne.

Mining.com. http://www.mining.com/several-casualties-after-tailings-dam-bursts-in-brazil/ (accessed 26 February 2017).

Napier F., D'Arcy B. and Jefferies C. (2008). A review of vehicle related metals and polycyclic aromatic hydrocarbons in the UK environment. *Desalination*, **226**, 143–150.

Nwankwo J. N. and Irrechukwu D. O. (1987). Problems of environmental pollution and control in the Nigerian petroleum industry. The proceedings of the international seminar on petroleum industry and the Nigerian environment, Nigerian National Petroleum Corporation (NNPC), Lagos, pp. 102–307.

Omodanisi E. O., Eludoyin A. O. and Salami A. T. (2014). A multi-perspective view of the effects of a pipeline explosion in Nigeria. *International Journal of Disaster Risk Reduction*, **7**, 68–77.

Osborn D., Every W. J. and Bull K. R. (1983). The toxicity of trialkyl lead compounds to birds. *Environmental Pollution (Series A)*, **31**, 261–275.

Pedersen K. E., Basu N., Letcher R., Greaves A. K., Sonne C., Dietz R. and Styrishave B. (2015). Brain region-specific perfluoroalkylated sulfonate (PFSA) and carboxylic acid (PFCA) accumulation and neurochemical biomarker responses in east Greenland polar bears (Ursus maritimus). *Environmental Research*, **138**, 22–31.

Porter E. (1973). Pollution in Four Industrialised Estuaries. The RCoEP, HMSO, London.

Pulles W., Heath R. G. M. and Howard M. (1996). A manual to assess and manage the impact of gold mining operations on surface water environment, WRC report No. TT79/96, South Africa.

Rodriguez A. G. P., Alberto J. A. N., Sanchez J. S., Rejon G. J. M., Gomez J. M. and Casillas T. A. D. V. (2015). Contamination by organochlorine pesticides in the aquifer of the Ring of Cenotes in Yucatan, Mexico. *Water and Environment Journal*, **29**, 140–150, CIWEM.

SFEI (2015). The Pulse of the Estuary: the state of Bay Water Quality, 2015 and 2065. SFEI Contribution no. 759. San Francisco Estuary Institute, Richmond, CA.

Singh A. K., Mondal G. C., Tewary B. K. and Sinha A. (2009). Major ion chemistry, solute acquisition processes and quality assessment of mine water in Damodar Valley coalfields, India. Proceedings of the International Minewater Conference, 19–23 October 2009, Pretoria, South Africa, ISBN 978-0-9802623-5-3.

Surveyor (1990). Toxic spill hits river, Surveyor, 15 March 1990, p. 5. ISSN 00 39 6303.

Tarras-Wahlberg N. H., Flachier A., Fredriksson G., Lane S., Lundberg B. and Sangfors O. (2000). Environmental impact of small-scale and Artisanal Gold Mining in Southern Ecuador. *Ambio*, **29**(8), 484–491.

Time (1969). The cities: the price of optimism. *Time Magazine*, 1 August.

USEPA (2014). http://iaspub.epa.gov/waters10/attains_nation_cy.control#total_assessed_waters (accessed March 14, 2014).

USEPA (2014). http://www.epa.gov/greenchemistry/presidential-gree-chemistry-challenge-award-recipients-industry-sector (accessed February 26, 2017).

USEPA (2017). http://www.epa.gov/greenchemistry/basics-green-chemistry#definition (accessed February 26, 2017).

Van Urk G. and Kerkum F. C. M. (1987). Ecological consequences of the fire at Sandoz. *$H_2O$*, **20**(3), 50–51. (in Dutch, English summary p. 49)

Villa S., Migliorati S., Serafina Monti G., Holoubek I. and Vighi M. (2016). Risk of POP mixtures on the Arctic food chain. *Environmental Toxicology and Chemistry*, doi: 10.1002/3671.

Wilson K. W., D'Arcy B. J. and Taylor S. (1988). The return of fish to the Mersey estuary. *Journal of Fish Biology*, **33**(Suppl A), 235–238.

Wilson K. W., Head P. C. and Jones P. (1986). Mersey Estuary (UK) Bird mortalities – causes, consequences and correctives. *Water Science and Technology*, **18**, 171–180.

Yoshida F., Hata A. and Tanegawa H. (1999). Itai-itai disease and the countermeasures against cadmium pollution by the Kamioka mine. *Environmental Economics and Policy Studies*, **2**, 215–229.

Younger P. L. (2000). Chapter 10: Iron. In: Diffuse Pollution Impacts, B. J. D'Arcy, J. B. Ellis, R. C. Ferrier, A. Jenkins and R. Dils (eds), Chartered Institution of Water and Environmental Management, London, pp. 95–104, ISBN 1 870752 46 5.

Younger P. L., Coulton R. H. and Frogatt E. C. (2005). The contribution of science to risk-based decision-making: lessons from the development of full-scale treatment measures for acidic mine waters at Wheal Jane, UK. *Science of the Total Environment*, **338**, 137–154.

Zagatto P. A., Gherardi-Goldstein E., Bertoletti E., Lombardi C. C., Martins M. H. R. B. and Ramos M. L. L. C. (1987). Bioassays with aquatic organisms: toxicity of water and sediment from Cubatao River Basin. *Water Science and Technology*, **19**(11), 95–106.

# Chapter 2

# Accidents and pollution: industry impacts

*Paul T. C. Harrison*[1]* *and Brian J. D'Arcy*[2]

*[1]PTCH Consultancy, Market Harborough LE16 9TU, UK*

*[2]independent environmental consultant, info@enviroexperience.co.uk*

**Corresponding author: paul@ptch.co.uk*

## 2.1 INTRODUCTION

Industrialisation has resulted in increased concentrations of pollutant substances in air, water and soil, mostly through slow and diffuse releases over prolonged periods of time. In addition there are occasional catastrophic releases of chemicals into the environment that can have dramatic, immediate and severe effects. The adverse consequences of any such releases are principally determined by the quantity released and the inherent toxicity and persistence of the substance and its propensity to bioaccumulate. Short and long term consequences may include both environmental degradation and adverse impacts on human health.

This paper takes a historical look at a few of the most significant impacts of industrial pollution on the environment, including catastrophic accidental releases involving sudden substantial releases of noxious substances, such as occurred at Bhopal and Seveso, as well as some examples of shocking, but less immediately obvious pollution situations and events. Better understanding of the risks of such exposures to both human health and the natural environment can enable informed decisions to be made about more controlled use of the substances or replacement by less hazardous alternative chemicals or processes. In Europe, chemical regulatory schemes such as REACH (registration, authorisation and restriction of chemicals)

© IWA Publishing 2017. Wealth Creation without Pollution: Designing for Industry, Ecobusiness Parks and Industrial Estates
Brian D'Arcy, Lee-Hyung Kim, Marla Maniquiz-Redillas
doi: 10.2166/9781780408330_25

aim to do just that, while the 'Seveso Directive', for example, is intended to reduce the risk and impact of major industrial accidents.

## 2.2 CATASTROPHIC EPISODES

There is a well documented history of 'catastrophic' pollution incidents over the past 40 years or so and some of the most infamous are outlined here.

### 2.2.1 Seveso, Italy

On 10th July 1976 there was a massive release of 2,3,7,8-tetrachlorodibenzo-p-dioxin (TCDD, 'dioxin') from a chemical plant in Seveso near Milan, Italy, which was manufacturing 2,4,5-trichlorophenol (2,4,5-TCP). A safety disc in a reaction vessel ruptured and a plume of chemicals containing 2,4,5-TCP and dioxin rose 30 to 50 m above the factory. The plume then grounded downwind of the factory, contaminating an area approximately 2 km long and 700 m wide. An estimated 3 to 16 kg of dioxin was released. Almost 28,00 people lived in the vicinity of the factory (Harrison, 2001).

Dioxin is both extremely toxic and chemically stable; it is known to cause the skin condition chloracne and affect foetal development. The population of Seveso was screened shortly after the accident and 176 individuals, mostly children, were found to have chloracne. A further round of medical screening six months later revealed a further 137 cases, but subsequent follow-ups showed the incidence to have lessened and symptoms improved. In addition to the chloracne, some neurological abnormalities were noted, including polyneuropathy, especially amongst people living in the most contaminated area of the city. Also, there was evidence of liver enlargement in about 8% of the population, which again was most noticeable amongst the most exposed population. However, there was no evidence of effects on the immune system or of chromosomal abnormalities or damage to foetuses. While no deaths were recorded as being due to the incident, a later analysis of cancer incidence in the exposed population found increased liver cancer and elevated incidences of leukaemia and other blood neoplasms in men and increased bone marrow cancers in women, as well as higher incidences of soft tissue tumours and non-Hodgkins lymphoma. Interestingly – and aligned with the fact that dioxin is an anti-oestrogen – breast cancer and endometrial cancer in women were reduced, and a changed sex ratio in the offspring of exposed men was reported (Harrison, 2001).

There were impacts too on the natural and agricultural environment. Many plants and herbivorous animals died (through oral uptake of contaminated vegetation) and large amounts of top-soil were removed from the site. Agricultural and horticultural activities were suspended over an area of approximately 1700 ha (Wipf & Schmid, 1983). It is known that TCDD becomes strongly absorbed to soil – hence its large-scale removal in contaminated areas – but the longer-term impacts of the incident on the natural environment remain largely unknown. TCDD is a proven endocrine disrupter with anti-oestrogenic properties, and hence reproductive effects in wildlife cannot be discounted (Institute for Environment and Health [IEH], 1995; Harrison *et al.* 1997). The devastation and public alarm engendered by the incident prompted the European Commission to adopt, in 1982, a Directive on the control of major accident hazards involving dangerous substances. Known as the Seveso Directive (presently amended and adopted as Directive 2012/18/EU, or 'Seveso III'), this is aimed at preventing such major chemical accidents.

### 2.2.2 Bhopal, India

The Bhopal disaster has been described as 'the worst industrial accident in history' (Broughton, 2005) and a true industrial catastrophe (MacKenzie, 2002).

This tragic event occurred on 3rd December 1984 and involved a catastrophic release of methyl isocyanate (MIC) from the Union Carbide factory that had been producing the insecticide carbaryl, MIC being one of the main ingredients. MIC itself was manufactured from monomethylamine and phosgene – a highly toxic gas that was produced on site by reacting chlorine and carbon monoxide. There was therefore at this site a variety of toxic substances in use or stored ready for use.

It is believed that the accident was caused by water entering a tank where 41 tonnes of MIC were being stored, causing a runaway chemical reaction. This resulted in rapid vaporisation of the tank contents, causing the safety valve to burst open. It remained open for about two hours, allowing MIC in liquid and vapour form – as well as other reaction products and contaminants – to escape into the immediate environment. Many people lived in the vicinity of the factory and as a consequence at least 2000 people died from exposure to the cocktail of extremely toxic substances. The most frequent symptoms amongst the individuals who survived were burning and watering eyes, coughing and vomiting. A great many more individuals continued to suffer physical and mental trauma as a result of this tragedy. In the eventual settlement between the company (Union Carbide Corporation) and the Indian Government, 3000 people were acknowledged to have died and 102,000 suffered permanent disabilities (Broughton, 2005), although the numbers have been disputed (e.g., Kumar, 2004). Subsequently, the Bhopal Gas Tragedy Relief and Rehabilitation Department reported that by the end of October 2003, compensation had been awarded to 554,895 people for injuries received, and to 15,310 survivors of those killed.

With such a catalogue of human deaths and disabilities, it is not surprising that impacts on wildlife and ecology have not been a major focus of the published literature. But the effects are likely to have been significant, given the human impact.

Groundwater contamination has certainly been recognised as one of the environmental consequences of the incident at Bhopal, a local aquifer having been contaminated (Fortun, 2001; Health and Safety Executive, 2004). The contamination of local aquifers by toxic organic chemicals and heavy metals has been linked with a failure to clean up completely after the incident (Broughton, 2005). Goodman (2009) reported that 25 years after the incident, the local groundwater – which provides a potable water supply for 15 communities – remained contaminated.

### 2.2.3 Sweizerhalle, Switzerland – the Rhine Incident

In 1986, a major fire at a pesticides factory near Basel, Switzerland, resulted in 30 tonnes of pesticides being washed into the River Rhine, carried in fire-water runoff. The fire broke out in a warehouse where a wide variety of chemicals, including mercury, were being stored. The factory was owned by Sandoz, one of the largest pesticide manufacturers in Switzerland. The inhabitants of Basel and the surrounding area on the border between France and Germany were told to stay indoors, as witnesses had reported a foul smell of rotten eggs and burning rubber. Fourteen people were admitted to hospital after inhaling the fumes (BBC, 1986). The runoff turned the river red and the toxic contaminants caused extensive fish mortalities in the tributary river and in the Rhine itself (Giger, 2009). One year after the event, a review paper written by the Head of Corporate Safety and Environmental Protection, Sandoz Group, reported that severe ecological damage had occurred over about 250 km of river, including the death of 'a great number of fish'. Eels (*Anguilla spp.*) predominated in this mortality. In the Upper Rhine region of Germany, downriver of Basel, dead fish were also found, with eels again worst affected. Toxic residues were detected in fish shortly after the incident, but not when measured one year later (Salzmann, 1987).

Lessons from the 1986 Rhine incident in relation to the storing of toxic and flammable substances included recognition of the importance of:

- The characteristics of buildings and their equipment (melting point, composition and formation of toxics on combustion, structural and functional integrity in the event of a fire involving the kinds of materials stored there)
- Storage density, storage volume and storage procedures
- Packaging materials and storage records
- Retention of fire-extinguishing water in case of fire.

The factory involved in the Rhine incident responded by addressing all the above points, building two catch basins of 15,000 $m^3$ and 2500 $m^3$ within the site, and modifying warehouse design to specify concrete fireproof walls, dividing walls and internal catchpits and sumps. Local fire officers recommended a storage volume of 3 $m^3$/tonne of warehoused material, based on their estimates of the volume of water normally dispersed into a fire (Salzmann, 1987).

During this incident the contaminated fire water passed down the river from Switzerland, through Germany and France, to the Netherlands, before discharging into the North Sea (Giger, 2009). This event set back a decade of intensive clean-up efforts for the Rhine, and had a major influence on European regulatory philosophy. The Rhine incident highlighted the trans-boundary nature of Europe's largest rivers and the need for river basin management, rather than attending simply to national interests. This ultimately led to the EU Water Framework Directive (Directive 2000/60/EC establishing a framework for Community action in the field of water policy).

### 2.2.4 Tianjin, China

Beginning on the night of 12th August 2015, a series of explosions of increasing magnitude occurred at a chemicals store at the Port of Tianjin, China. The store was operated by Ruihai Logistics, a privately held company established in 2011, handling hazardous chemicals including flammable and corrosive substances, oxidising agents and other toxic chemicals. The 46,000 $m^2$ site contained multiple warehouses for hazardous goods, a fire pump and a fire pond. The business was operating in breach of several requirements relating to storage quantities, licences and safety regulations.

The first reported fire was apparently due to auto-ignition of nitrocellulose which had become warm; the fire then led to a series of explosions, the biggest of which involved 800 tonnes of ammonium nitrate. The last explosions continued into 15th August. There was some uncertainty about the chemicals stored at the site, but there were over 40 kinds of hazardous substances present, totalling about 3000 tonnes (Zeng, 2015), including calcium carbide as well as nitrates and at least 700 tonnes of toxic sodium cyanide (about 10 times the legal limit there). Firefighters were unaware of the presence of calcium carbide, which releases the flammable gas acetylene on contact with water, hence contributing to the fires and explosions.

A total of 173 people died in the disaster, with 797 non-fatal injuries. Over one thousand firefighters attended the incident, of whom 95 died – the biggest loss of life of first-response staff in China since the founding of the People's Republic of China in 1947. As well as destroying the site, damage to nearby buildings and property was serious, including over 8000 new cars. Groundwater contamination, for example by cyanide, was a major concern. Fish mortality occurred when the first rain fell after the incident, washing contaminating chemicals either from the atmosphere or from the surfaces of the area affected by dry deposition, or both. Local officials suggested the mortality (of sticklebacks, one of the most resilient fish species, able to tolerate poor quality conditions) could have been due to low oxygen concentrations (it was August, following a dry period when this would be likely in an urban watercourse). People reported white chemical foam covering the streets when it rained, and burning sensations and skin rashes when in contact with rain droplets (China Daily, 2015a; Phillips, 2015; Varghese, 2015).

## 2.3 CHRONIC POLLUTION IMPACTS

In addition to the catastrophic events described above, a number of serious *chronic* pollution episodes have occurred over recent history. Some of the most severe and significant events are detailed here.

### 2.3.1 Hinkley, California

The groundwater contamination incident at Hinkley, in the Mojave Desert, California, was made famous by the Hollywood movie *Erin Brokovic*, released in 2000. The focus of the litigation made famous in the film, was the risks to the health of a local population drinking from wells affected by chromium leaking into the local groundwater. Hexavalent chromium had been deliberately added by Pacific Gas and Electric Company (who operated a nearby natural gas compressor station) to water in cooling towers to prevent rusting of machinery parts. This water was stored between use in unlined ponds (California Environmental Protection Agency [EPA], 2016). From an industrial good practice point of view, and indeed with regard to business efficiency, the Hinkley case presents an example – for new industries anywhere – of a demonstrably poor environmental philosophy. Leaving aside possible health risks to groundwater abstractors and the likely long-term ruination of a potable water resource for local inhabitants, it is difficult to comprehend how a company could store the chromium contaminated water in *unlined pits* in an arid region. This is not an example of an 'accident' or a technically challenging effluent management problem. The cost of basic pollution prevention measures would have been a tiny fraction of the eventual legal costs of the action taken against the company – more than $330 million (KPCC, 2012).

Chronic groundwater contamination resulted from the start of the practice in 1952 until stopped in 1966. The plume of contaminated groundwater was approximately 3.2 km long and 1.6 km wide, though the total area of groundwater affected was estimated to be at least 12 km by 3 km. Average hexavalent chromium levels in Hinkley were recorded as 1.19 ppb, with an estimated peak of 20 ppb. The current federal (EPA) drinking water standard for total chromium is 100 ppb, but is under review. The California EPA has established a Public Health Goal (PHG) of 0.02 ppb for hexavalent chromium based on avoidance of potential carcinogenic effects (California EPA, 2016). The issue is still ongoing with the remaining local community in the town, and continuing measures are being sought from the company by the environmental regulators, encouraged by local residents groups (see actions required by order of Lahontan Regional Water Quality Control Board, www.swrcb.ca.gov/nwqrb6/waterissue, and the website for the Hinkley Community by the Independent Review Panel (IRP) Manager www.hinkleygroundwater.com/category/news-articles).

## 2.3.2 Love Canal, USA

The reporting of chemical odours in the basements of homes in the Love Canal neighbourhood of Niagara Falls, USA, led to toxicological investigations that had prolonged repercussions for the practice and regulation of waste disposal in the USA, and indeed the rest of the world.

Love Canal was a waste disposal site containing both municipal and industrial waste that had been deposited over a period of more than 30 years, up to 1953. According to reports, the site had been used for the disposal of some 22,000 tons of chemical waste by the Hooker Chemical Company. Homes were subsequently built on the site during the 1960s and leachates began to be detected in the late 1960s. Substances detected in the organic phase of the leachates included dibenzofurans and dioxins, both of which have significant toxic properties posing the risk of possible immunologic, carcinogenic and teratogenic effects in exposed individuals. The episode resulted in significant fears of ill-health and much psychological stress amongst the residents. Limited follow up of the local population identified low birth weights in the offspring of Love Canal residents, but no definite causal link was established in relation to cancer incidence in the area (Harrison, 2001).

The Love Canal incident provides an example where considerable public anxiety and stress can result from the identification of a potential severe toxic hazard. Indeed, it is clear that the psychological and other consequences of such incidents might well outweigh the actual toxic effects of exposure, and are no less important. Concern specifically about the possible health impacts of landfill sites resurfaced in later years with the suggestion of an association between congenital malformations and residence close to landfill sites. While there is no evidence for a causal relationship for such findings, the question of possible health impacts from waste disposal remains pertinent.

### 2.3.3 Minamata and Niigata, Japan

The pollution episodes at Minamata and Niigata in Japan represent probably the best known examples of endemic disease caused by environmental exposure to organic mercury. The condition – now referred to as 'Minamata disease' – was first noted at the end of 1953 when an unusual neurological condition began to affect villagers living on Minamata Bay on the south-west coast of Kyushu. Both sexes and all ages of inhabitant were affected, presenting with a range of symptoms relating to the peripheral and central nervous system. Many patients became disabled and bedridden, and about 40% died. The disorder was associated with consumption of fish and shellfish caught in the bay, which was contaminated with mercury from effluent released from the Chisso Corporation's chemical factory, manufacturing vinyl chloride using mercuric chloride as a catalyst (Harada, 1995). It was claimed at the time that the inorganic mercury was methylated by microorganisms living in the bay, which then entered the food chain contaminating the marine food species. However, it is now considered likely that the mercury was directly released in its organic form. While cat, dog, pig, and human deaths continued for 36 years, the government and company did little to prevent the pollution. As of March 2001, 2265 victims had been officially recognised as having Minamata disease (1784 of whom had died) and over 10,000 had received financial compensation from Chisso. By 2004, Chisso Corporation had paid $86 million in compensation, and in the same year was ordered to clean up its contamination. On March 29, 2010, a settlement was reached to compensate as-yet uncertified victims (Harrison, 2001; Allen & Burns, 2009; Wikipedia, 2016).

Remarkably, a second outbreak of methyl mercury poisoning occurred in Japan in 1965, affecting hundreds more individuals living in the Niigata Prefecture. This episode followed pollution of the Agano river by industrial effluent and, again, the consumption by the local population of fish in which the mercury had bioconcentrated (Harrison, 2001; Allen & Burns, 2009). Whilst the focus of concern in both these cases is human health, it is certain that diseases associated with oral exposure to methyl mercury would have adversely affected all top predators in the region.

### 2.3.4 Toyama Prefecture, Japan

Prolonged environmental exposure to cadmium in the Toyama Prefecture in Japan resulted in a severe outbreak of itai-itai disease, which means literally 'it hurts'. First reported in 1955, it occurred in the population in an area downstream from a mine on the Juntsu River. The condition was almost entirely confined to elderly women who had borne several children, and was characterised by severe bone pain, bone softening, fractures, waddling gait and renal impairment. The river water that was used to irrigate crops was frequently contaminated by outpourings from the mine, which contained zinc, lead and cadmium. Levels of cadmium in locally harvested rice were shown to be about 10 times higher than normal, leading to the conclusion that cadmium exposure was responsible for the disease. However, calcium and vitamin D deficiencies in the local population were almost certainly important contributory factors (Harrison, 2001). Local wildlife is likely to have also suffered the consequences of chronic metal poisoning from exposure to the untreated mine effluent.

## 2.4 DISCUSSION

When human health risks are involved, there are two powerful driving forces at play: justifiable public concerns in the face of contamination by known toxicants, and the industrialists' fear of multi-million-dollar litigation for health impacts on communities. Investigations and remedial actions are easier when human health is not the primary concern – for example where impacts are on aquatic life in receiving waters. Invertebrates don't sue for compensation, and no-one supposes sub-lethal effects observed in fish after a chemical spill are more likely to be psychosomatic. For human populations – where people's health is arguably the most important concern regarding environmental exposures – there is often a real risk of literally adding insult to injury in pursuit of an almost impossibly high level of proof of environmental damage and liability.

Several of the examples in this paper demonstrate the problems of investigating pollution incidents *in situ*: they are not scientific research projects with proper controls, or carefully run trials dosing the suspected chemicals of concern, isolating other possible variables and influences. For example, it is obviously neither practical nor ethical to conduct blind trials using doses of selected chemicals on one group of volunteers whilst given identical looking substances to a control group, neither group knowing what they've been given – and running such trials for months or years to simulate chronic exposure. An ongoing investigation is the opposite of a planned study and there are some important 'unknowns': Is it a one-off sudden leak of something, or a hitherto undetected chronic problem? Is it actually a pollution problem, or something else? Sometimes it is by no means clear which are the chemicals to be investigated in an effluent 'brew' of product, intermediates, process side reactants, raw materials and the spectrum of forms in between. The difficulty of selecting the right chemical to study is, however, just one aspect. What about synergistic effects of mixtures? The latter would add greatly to the complexity and costs of a thorough scientific trial (and, it must be stressed, a field investigation into an ongoing and varying environmental pollution episode is not a scientific trial evaluating health impacts on populations or even physical dispersion/degradation and uptake risks).

In the 1990s, the then UK Environment Department (DETR) commissioned an important report into methods to be adopted in the event of a chemical accident (AEA Technology Environment, 1999). The objective was to ensure that appropriate environmental samples and monitoring of environmental impacts is planned for and undertaken in the event of a chemical pollution incident. Such information would be very useful in analysing the consequences of – and mitigating the effects of – environmental degradation following such an event. It is not known how such guidance has been adopted or is followed, but to properly understand the environmental impacts of a major pollution incident this is absolutely essential.

Several of the cases reported here involved considerable expenditure related to clean-up costs and subsequent legal settlements – often far exceeding the cost of installing protective measures in the first place. Table 2.1 provides a summary of impacts and costs of some of the incidents described here.

Several of the accounts of major accidents or incidents outlined above reported psychological effects on local communities – including residents nearby, workers and officials. Impacts on public and business confidence do not receive as much attention as chemical contamination, but perhaps deserve more consideration in designing and costing safety and accident prevention measures at a plant. Salzmann (1987), writing for Sandoz, highlighted the psychological impacts on residents around Basel who, although not adversely affected by the chemicals released in the fire at the Sandoz plant, were nonetheless in shock that such an event could happen at 'their' local factory. The loss of public confidence in the company was felt to be a major ongoing concern for the business.

Although this paper is primarily concerned with 'acute' major chemical incidents, two of the severe classic pollution episodes reviewed here went on for many years. Sadly, it is very likely that in developing countries seeking to industrialise their economies, examples of equally poor environmental management, destroying resources for others and the health and viability of local communities, are still occurring. Continuous pollution is inefficient business – loss of resources for the business and destruction of resources for others in the locality.

## 2.5 CONCLUSIONS

Impacts of industrial accidents involving catastrophic or severe chronic pollutant releases include:

- Health impacts on those seeking to contain an incident at the site
- Health impacts on the local human population

**Table 2.1** Summary of impacts and costs to industry of some of the pollution incidents.

| Company | Problem | Contributory Cause/s | Subsequent Legal Settlement Cost in $US | Impact |
|---|---|---|---|---|
| Union Carbide[1,2] Bhopal plant, India | Leak from pesticide plant & explosion | Production continued whilst decommissioning plant and shut down safety features[2] | $470 million | >3800 people killed[2] |
| Sandoz[3,4], Rhine Incident, Switzerland | Fire in pesticide store | Unable to contain volume of fire-water; quantity of toxic materials stored on premises | | Fish mortalities & other ecological impacts Loss of confidence in the business |
| Pacific Gas & Electric Co.[5] Hinkley, California, USA | Storing toxic chemicals in solution in unlined vessel in permeable soils supplying groundwater | Denial of risks | In excess of $333 million | Chronic exposure to $Cr_6$ contamination in drinking water supplies |
| Chisso Corporation[6], Minamata, Japan | Effluent untreated, then inadequately treated, then diverted to another waterbody | Denial of a problem, perhaps initially ignorance of pollution prevention techniques | $3.4 million compensation to victims awarded in 1973 court action, rising to a total $86 m by 2004 | Severe health impacts and hundreds of deaths, plus congenital impacts for decades |
| Ruihai Logistics[7], Tianjin, China | Explosion | Mix of toxic and explosive chemicals in excessive quantities | $360,669 paid by govt. to each of the families of firefighters killed | 173 dead, extensive damage to property, human health risk, groundwater pollution risk |

*References:* [1]Goodman (2009), [2]Broughton (2005), [3]Salzmann (1987), [4]Giger (2009), [5]KPCC (2012), [6]Allen and Burns (2009), [7]China Daily (2015b).

- Environmental damage (e.g., polluted groundwater, contaminated ecosystems, fish mortalities)
- Destruction of resources for others, thereby adversely affecting the viability of local communities (local fisheries, potable water supplies, abstraction for other businesses downstream and hence employment risks)
- Loss of confidence in the local factory where the incident happened
- Loss of confidence/trust in the company responsible
- Loss of confidence in the industry.

Causes of major accidents vary, but include:

- Poor management
- Inadequate inspection and maintenance procedures
- Inadequate priority given to a precautionary approach to risks and human health and safety, and environmental protection, by the company responsible
- Inadequate contingency planning for emergencies, especially lack of provision for managing fires and fire-water
- Failure to understand that the cost of prevention measures is usually trivial compared to the cost of impacts when accidents or other serious pollution episodes occur
- In a few examples, ill-informed dogma in relation to environmental issues; allowing common sense to be over-ruled by a reluctance to sanction funds for basic measures at relatively trivial cost.

Insurance policies can cover costs for restoration of property, but cannot restore human fatalities. Investment in sensible precautions and safe practices is the only appropriate insurance policy for a business with risks to people. This paper serves as a reminder why a measured precautionary approach by government, communities and especially industry is sensible and appropriate. A few of the examples here underline why a proper regulatory regime is very necessary, certainly for the good of the local populations and environment but also in many instances in the interests of sustained production at the industrial premises themselves. The disaster at Tianjin in China is a topical example; the question is, how many more such incidents will happen even though basic prevention and risk management procedures are usually neither expensive nor technically difficult?

## REFERENCES

AEA Technology Environment (1999). Environmental Sampling after a Chemical Accident. Stationery Office: ISBN 0117534951, 9780117534957. 354pp.

Allen K. and Burns C. (2009). Minamata disease. Boston University, Boston. http://www.eoearth.org/view/article/154624/

BBC (1986). 1986: Chemical spill turns Rhine red. http://news.bbc.co.uk/onthisday/hi/dates/stories/november/1/newsid_4679000/4679789.stm

Broughton E. (2005). The Bhopal disaster and its aftermath: a review. *Environmental Health: A Global Access Science Source*, **4**, 6. Available from http://www.ehjournal.net/content/4/1/6

California EPA (2016). http://www.swrcb.ca.gov/rwqcb6/water_issues/projects/pge/index.shtml; http://www.swrcb.ca.gov/rwqcb6/water_issues/projects/pge/index.shtml and http://oehha.ca.gov/water/phg/072911Cr6PHG.html

China Daily (2015a). Tianjin detects harmful air pollutants, Beijing stays clear. 13 August 2015. Retrieved 14 August 2015.

China Daily (2015b). Tianjin announces compensation for martyred firefighters families. Chinadaily.com.cn. Retrieved 14 August 2015.

Fortun K. (2001). Advocacy after Bhopal. University of Chicago Press, Chicago, p. 259.

Giger W. (2009). The Rhine red, the fish dead – the 1986 Schweizerhalle disaster, a retrospect and long-term assessment. *Environment Science and Pollution Research*, **16**(Suppl 1), S98–S111.

Goodman S. (2009). Poisoned Water Haunts Bhopal 25 Years after Chemical Accident. http://www.scientificamerican.com/article/groundwater-contaminatioon-india-pesticide-factory

Harada M. (1995). Minamata disease: methylmercury poisoning in Japan caused by environmental pollution. *Critical Reviews in Toxicology*, **25**(1), 1–24.

Harrison P. T. C. (2001). Health effects of environmental chemicals. In: Pollution Causes, Effects and Control, R. M. Harrison (ed.), 4th edn. Royal Society of Chemistry, Cambridge, UK, pp. 500–523.

Harrison P. T. C., Holmes P. & Humfrey C. D. N. (1997). Reproductive health in humans and wildlife: are adverse effects trends associated with environmental chemical exposure? *Science of the Total Environment*, **205**, 97–106.

Health and Safety Executive (2004). Accident Summary, Union Carbide India Limited, Bhopal, India: December 3, 1984. In: Hazardous Installations Directorate. HSDA.

IEH (1995). IEH assessment on environmental oestrogens: consequences to Human Health and Wildlife (Assessment A1). In: P. T. C. Harrison, C. D. N. Humfrey, M. Litchfield, D. Peakall and L. Shuker (written and edited), MRC Institute for Environment and Health, Leicester, UK. Available at: http://www.iehconsulting.co.uk

KPCC (2012). PG&E makes $3.6 million settlement with Hinkley, its second in 20 years. http://www.scpr.org/news/2012/03/16/31664/pge-makes-36-million-settlement-hinkley-its-second/

Kumar S. (2004). Victims of gas leak in Bhopal seek redress on compensation. *British Medical Journal (BMJ)*, **329**(7462), 366.

Li J., Huang, K. L. and Jun, M. (2015). Mystery white foam and burns reported as rain falls at Tianjin blast site sparking fears of toxic reactions. South China Morning Post, 26 August 201, China.

Mackenzie D. (2002). Fresh evidence on Bhopal disaster. *New Scientist*, **19**(1).

Phillips T. (2015). Chinese authorities investigate mass fish death near Tianjin explosion site. *The Guardian*, 20th August 2015. Retrieved 24 August 2015.

Salzmann J. J. (1987). "Schweizerhalle" and its consequences. In: Fires Involving Pesticides – The Rhine Incident. Proceedings of a Seminar on 25/26 November 1987. British Agrochemicals Association.

The Guardian (2015). Tianjin explosion: China sets final death toll at 173, ending search for survivors. *The Guardian*. Associated Press. 12 September 2015.

Varghese J. (2015). Tianjin Aftermath: 'Chemical Rain' leaves streets foaming, residents complain about rashes. *The International Business Times*, 18 August 2015. http://www.ibtimes.co.in/tianjin-aftermath-chemical-rain-leaves-streets-foaming-residents-complain-about-rashes-643377 (Retrieved 18 April 2016)

Wikipedia (2016). Minamata disease. https://en.wikipedia.org/wiki/Minamata_disease

Wipf H. K. and Schmid J. (1983). Seveso – an environmental assessment. In: Human and Environmental Risks of Chlorinated Dioxins and Related Compounds, R. E. Tucker, A. L. Young, A. P. Gray (eds), Plenum Press, New York, NY.

Zeng V. (2015). 3,000 Tonnes of Dangerous Chemicals were Stored at Tianjin Explosion Site, Say Police. Hong Kong Free Press, 18th August 2015.

# Chapter 3

# Industrial estates as sources of water pollution

*Lee-Hyung Kim[1], Brian J. D'Arcy[2]*, Marta Ibanez[3] and Marla Maniquiz-Redillas[1]*

*[1]Kongju National University*

*[2]Independent consultant*

*[3]BEAR Scotland*

**Corresponding author: brian@enviroexperience.co.uk*

## 3.1 INTRODUCTION

### 3.1.1 Definitions

Chambers 21st Century Dictionary (1999 edition) defines an **industrial estate**

> 'an area in a town which is developed for industry and business. Also called **trading estate**. *N American equivalent* **industrial park**.'

The Oxford Dictionary for the Business World (1993) Oxford University Press, also equates industrial estate with trading estate, and states that it is:

> 'a specifically designed industrial and commercial area, usually sited some distance from residential areas.'

Unfortunately, the specific needs of industry to generate wealth without causing pollution, have not until recently been sufficiently addressed. In terms of pollution risks, an industrial estate is an agglomeration of different premises involving loading and unloading of materials, processing in various ways a variety of materials, and often *ad hoc* cleaning and washing activities, storage of a variety of materials indoors

© IWA Publishing 2017. Wealth Creation without Pollution: Designing for Industry, Ecobusiness Parks and Industrial Estates
Brian D'Arcy, Lee-Hyung Kim, Marla Maniquiz-Redillas
doi: 10.2166/9781780408330_37

and outdoors, and typically involving substantial vehicle traffic, truck fleets and staff cars. A large manufacturing site owned by one company (e.g., a brewery or a car plant) is an industrial estate in terms of diverse pollution risks for surface water, as much as an assortment of companies in an industrial/commercial zoned area. This paper indicates the polluting nature of industrial estates for the water environment, in the absence of sustainable drainage and pollution prevention strategies for the examples cited or described.

### 3.1.2 Pollution sources and importance

It may be a surprise to many, but in any large industrial estates usually only a handful of the industries have recognised, authorised process effluent discharges. Most industrial units on an estate will be dry processes such as packaging, storage and distribution, machining and other engineering works, construction, logistics, metal finishing, assembly, and more. The absence of direct effluent discharges however, and a good safety record for pollution accidents, is not enough to prevent pollution from industry. Many industrial estates are very large, with numerous companies of all kinds spread over extensive areas. All have surface water drainage, and stormwater runoff is often contaminated by many diffuse sources. Pollution of the water environment from such extensive industrial and commercial zones is diffuse in nature, which is not to say that there are not hotspots where pollutants occur in greater quantities, and some activities present far more risk than others. Some are *ad hoc* trade effluents, such as mobile steam cleaning units, which are very difficult to regulate. There is a complete spectrum of pollution incidents and accidents from the individually trivial but often regular (e.g., drips of oil, or chemicals gradually accumulating on roofs, the ground and in drains) to inadequate disposal of potentially polluting wastes, and large-scale irregular spills. That direct industrial/commercial activity overlies background contamination arising from combustion of fuel, power generation, heating/air conditioning, use of commercial vehicles and machinery, and includes roof runoff contaminated by extractor fans in machine workshops discharging very fine particulate pollutants onto roof areas. Diffuse pollution hotspots occur when polluting activities result in significant contamination at particular premises or areas of ground or roofs, and pollutants are liable to be mobilised by rainfall and washed into drainage networks or otherwise into the aquatic environment. Industrial districts are often a worst case land-use for diffuse source pollution. Table 3.1 compares unit pollution loads for different land-uses and indicates that industrial and commercial areas can be important for several of the relatively small spectrum of contaminants measured in the studies cited.

In terms of characterising industrial estate pollution, it can be difficult to separate the effects of trade effluent discharges on an estate from diffuse pollution (rainfall mobilised wash off or seepage associated with chronic contamination or major accidents). Using water quality data (chemical and biological) interpreted by staff with expert local knowledge, the Scottish Environment Protection Agency

(SEPA), assessed the impacts of industrial estates as causes of water pollution in 1999 and compared them as diffuse pollution sources with other urban and with trade effluent discharge impacts (Table 3.2). The pattern was similar to that which might be expected from the literature data in Tables 3.1(a) and (b).

**Table 3.1(a)** Single study of unit pollutant loads from homogeneous urban watersheds in Milwaukee, USA.

| Land Uses (kg/km²/day) | TSS | TP | Pb |
|---|---|---|---|
| Highways | 268.22 | 0.28 | 1.36 |
| **Industrial areas** | **262.19** | **0.41** | **0.74** |
| **Commercials areas** | **262.19** | **0.41** | **0.74** |
| Parking lots | 142.11 | 0.21 | 0.26 |
| High density residential areas | 133.42 | 0.31 | 0.25 |
| Medium density residential areas | 59.18 | 0.16 | 0.07 |
| Low density residential areas | 3.01 | 0.01 | 0.00 |
| Parks | 0.82 | 0.01 | 0.00 |

*Source:* Bannerman *et al.* (1983).

**Table 3.1(b)** Comparison of unit pollutant loads (literature 1983–2012).

| References | Land Uses | TSS (kg/km²/day) | COD (kg/km²/day) | TN (kg/km²/day) | TP (kg/km²/day) | Pb (kg/km²/day) |
|---|---|---|---|---|---|---|
| Kim *et al.* (2012) | Industrial | 143.86 | 52.45 | 6.65 | 1.75 | 0.10 |
| MOE (1999) | Urban | | | 13.69 | 2.10 | |
| Bannerman *et al.* (1983) | Industrial | 262.19 | | | 0.41 | 0.74 |
| Lee *et al.* (2008) | Highway | 399.50 | 356.30 | 12.29 | 2.46 | |
| Go *et al.* (2009) | Road | 580.13 | 331.17 | 14.68 | 1.43 | |
| NFWMD (1994) | Industrial | 213.73 | | 3.05 | 1.46 | |
| Novotny *et al.* (1997) | Industrial | 262.25 | | | 0.41 | 0.74 |

Current water quality data for industrial estate surface water drainage discharges to watercourses is limited (the problem was well known and thought by many to be intractable), but spot samples were routinely collected in Scotland in the 1990s, as well as the biological data used in conjunction with that, to derive the impacts in Table 3.2. Table 3.3 is a summary of some of the industrial estate water quality evidence. It is almost certainly as relevant now as it was then, in many countries, wherever comparable industrial activities occur on industrial areas with hard standing and runoff. As well as showing the variable quality of industrial estate surface water drainage, Table 3.3 gives an idea of the number of industrial estates in the study river catchment (the River Forth in East Scotland); a widespread pollution source, wherever there are settlements.

**Table 3.2** Importance of industrial estates as causes of pollution in Scotland.

| Pollution Sources | Rivers km Downgraded | Comments |
|---|---|---|
| *Urban* (all categories, including industrial estates) | 500 km | Often impacts most evident in small watercourses |
| *Industrial estates*, including airports | 168 + 47 km | Included several of the poorest quality watercourses in Scotland |
| *Trade effluents (for comparison)* Including textiles, food & drink, paper, electronics and petrochemicals | 66 km | Waste minimisation including resource recovery, together with effective effluent treatment technology reduced detrimental impacts |

*Source:* SEPA (1999).

**Table 3.3** Quality of surface water drainage from industrial estates in the catchment of the River Forth, Scotland (mean and max data from spot samples, 22 locations, 1990–1993[1]).

| EQS, Environmental Quality Standard for a Scottish River[2] | 25 mg/l (90%ile) | 4 mg/l (90%ile) | 0.78 mg/l (95%ile) |
|---|---|---|---|
| **Surface Water Outfall, SWO or Watercourse** | **Total Suspended Solids, mg/l Mean (max.)** | **$BOD_5$, mg/l Mean (max.)** | **Ammoniacal Nitrogen, mg/l Mean (max.)** |
| Queensway | 10 (70) | 6 (44) | 0.2 (1.0) |
| Eastfield | 14 (54) | 21 (230) | 1.1 (10.6) |
| Bankhead | 58 (169) | 6 (15) | 0.9 (2.5) |
| Nether Stenton | 39 (279) | 21 (180) | 0.6 (2.4) |
| Southfield | 11 (49) | 4 (11) | 0.2 (0.4) |
| *Lower Bighty* | *13 (70)* | *6 (29)* | *0.2 (0.7)* |
| *Lyne Burn d/s Elgin St.* | *60 (650)* | *4 (23)* | *1.0 (9.4)* |
| Lochlands Industrial estate | 183 (1730) | 29 (190) | 0.7 (3.5) |
| Cowie Industrial estate | 22 (199) | 12 (152) | 1.9 (7.5) |
| Cumbernauld No. 8A | 80 (216) | 28 (126) | 0.2 (0.4) |
| Cumbernauld No. 8 | 56 (84) | 14 (38) | 3.4 (7.3) |
| Cumbernauld No. 1 | 27 (129) | 31 (154) | 0.2 (0.6) |
| *Red Burn d/s Cumbernauld SWOS* | *15 (158)* | *3 (16)* | *0.1 (1.2)* |
| Old Liston industrial estate | 12 (29) | 7 (20) | 0.4 (1.6) |
| Gyle industrial estate | 58 (183) | 5 (14) | 0.6 (0.8) |
| East Mains industrial estate | 225 (3920) | 9 (80) | 0.5 (1.7) |
| Houston industrial estate | 36 (252) | 12 (47) | 5.5 (32) |
| Houston industrial estate north | 34 (125) | 3 (9) | 0.3 (1.8) |
| Deans industrial estate | 23 (134) | 6 (14) | 0.1 (0.2) |
| Brucefield industrial estate | 29 (122) | 4 (15) | 0.7 (2.7) |
| Kirkton Campus | 27 (78) | 17 (63) | 1.9 (7.7) |
| South Deans industrial estate | 7 (29) | 7 (45) | 2.1 (15.7) |

[1]Table adapted from D'Arcy and Bayes (1995).
[2]EQS values from Appendices in D'Arcy *et al.* (2000).

Table 3.3 also indicates the severity of contamination likely to occur, but cannot indicate whether this is as a result of an incident, or a storm event. The mean data from a set of spot samples is probably biased to dry weather sampling, but it does suggest evidence that an industrial estate is a pollution risk, especially on small watercourses with little catchment contribution from cleaner land-uses.

## 3.2 CASE STUDY (1) NW ENGLAND & WALES

### 3.2.1 Sources and pathways – Industrial estate surveys in Merseyside

The problems are not new. The importance of industrial estates as intractable chronic pollution problems was recognised by the former North West Water Authority in a report in 1984, investigating water quality in the River Mersey and tributaries and the adjoining catchment of the River Douglas, in NW England (D'Arcy *et al.* 1984). It was a detailed study and reported here since it is probably still very relevant in most countries where waste storage and diffuse pollution legislation has not been established and enforced. Table 3.4 summarises water quality and impacts for one of the most severely impacted streams in that intensive survey (Kirkby Brook).

**Table 3.4** Organic pollutants in drainage from Kirkby Industrial Estate (Kirkby 72 inch surface water outfall) in 1984, and impacts on stream quality.

| Kirkby 72" SWO into Kirkby Brook | $BOD_5$ | COD |
|---|---|---|
| Mean concentration | 138 mg/l | 354 mg/l |
| Max. concentration | 1120 mg/l | 1790 mg/l |
| Min concentration | 12 mg/l | 75 mg/l |
| Estimated load | 152 kg/d | 539 kg/d |
| **Kirkby Brook** | **Watercourse originates within estate** | **Classification below 72" SWO** |
| Water quality class (1 is best and 4 is poorest) | – | Class 4 |

*Source:* D'Arcy *et al.* (1984).

It is important to note that the data in Table 3.4 above was compiled as part of a programme of intensive (weekly) visits to potential sources on the estate in a systematic effort to resolve pollution problems. The work demonstrated that diffuse pollution is not able to be solved by the approach taken then of just putting bund walls around oil tanks and diverting contaminated areas to the foul sewer.

Twenty-two industrial estates were identified for assessment by biological sampling above and below their surface water drainage discharges into the receiving watercourses, as exemplified above in Table 3.4 for Kirkby Industrial Estate. In most instances there was a deterioration in quality below the estate, or the quality was poor and the watercourse originated close to the estate. The survey results are summarised in Table 3.5.

**Table 3.5** River quality below industrial estates in Mersey-Douglas catchments, 1984 (class 1 is good, 4 is bad).

| River Water Quality Class Below Surface Water Outfall In River | No. of Watercourses | Comments |
|---|---|---|
| Class 1 downstream (d/s) | None | Only 1 site where upstream (u/s) quality was 1 (d/s was 3) |
| Class 2 downstream | 2 | Aesthetic issues (oil) in both cases, but no class change |
| Class 3 downstream | 10 | 1 site had improved from class 4 previous year |
| Class 4 downstream | 7 | 1 site had deteriorated from previous year |
| All drainage diverted to foul sewer prior to survey | 1 | Compromise action to try to stop river pollution |
| No data obtained | 2 | Watercourse unsuitable for biological sampling |

*Source:* Summarised from D'Arcy *et al.* (1984).

The 1984 biological survey (Table 3.5) was undertaken following a period of several years of intensive investigations to try to achieve improvements (Table 3.6). The biologists remarked that although the rivers and streams remained in the lowest pollution class category, there had been a very clear improvement in visual appearance of many of the watercourses. But gaining class improvements obviously required more than just 'housekeeping' measures such as storage bunds and clean-up actions.

**Table 3.6.** Summary of the findings for 6 different estates.

| Industrial Estate | No. Years of Investigations | No. Sources Contamination | Pollution Issues in Watercourses |
|---|---|---|---|
| 1 | 2 | 22 | Oil, toxic metals, foam, suspension, paint |
| 2 | 3 | 23 | Oil, foul smell, sewage fungus, turbidity, acid |
| 3 | 1 | 27 | Oil, discolouration, toxic metals |
| 4 | Several | Many | Oil, dyes, turbidity |
| 5 | 2 | 24 | Oil, cyanide, acid, smell, discolouration |
| 6 | 2 | 7 | Oil, discolouration |

Table 3.7 summarises the contaminants and pathways for pollution identified in the Mersey-Douglas investigations. Similar problems have been found in similar investigations elsewhere (D'Arcy & Bayes, 1995) and there is little reason to expect anything different in other countries unless pollution prevention measures and sustainable drainage systems (SUDS) (CIRIA, 2015) or urban best management practices (BMPs) (United States Environmental Protection Agency [USEPA], 1993) have been provided. In the years since that baseline study, in the UK there has been a succession of new laws and powers to control

and prevent pollution, especially associated with the EU Water Framework Directive and the parallel Waste Framework Directive. A repeat of the 1984 survey would be useful to assess the effectiveness of those measures, since waste storage outdoors has been changed as a result and should be reflected in water quality improvements with lessons for mitigation plans elsewhere and catchment management programmes. The basis of river quality classification has also changed, but comparisons could still be made since the former methodology is published and still available.

**Table 3.7** Causes of surface water contamination identified in surveys of industrial estates.

| Type of Industry | Example Pollutant | Pathway for Surface Water Contamination |
|---|---|---|
| Most industrial premises | Fuel oils<br>Gas oil<br>Diesel | Unbunded tanks<br>Open valves on bund drain<br>Filler points outside bund areas<br>Corroded feed pipes from stock tanks<br>Corroded steam pipes from heavy fuel oil |
| Light & heavy engineering (e.g., motor vehicle assembly, zip factory, metal finishing) | Lubricating oils<br>Hydraulic oil<br>Cutting oil | Waste swarf skips – spillages<br>Surface water gullies inside factory<br>Road gulleys close to wastes |
| Cable manufacture | Cadmium, copper | Dust (from air filters) stored in open drums outdoors (open to rainwater leaching metals) |
| Breweries | Yeast, beers, fob (foam), waste beer | Spills from road tankers at loading bays, pressure release valves, can crushing plant in open yard |
| Breweries, distilleries, animal feed industry, granaries | Fermenting grain | Wind-blown grain from unenclosed off-loading/handling areas, carried by wind/rain into drainage system |
| Basic chemicals manufacture, chemical industry | Caustic alkali, acids, sodium bisulphide | Unbunded tanks, filler points, leaks, spills, corrosion |
| Food processing | Molasses, flour, sugar, other highly degradable materials | Pipe bursts and valve failures, leaks, brushing out spilled materials from burst containers in vans/wagons outdoors |
| Industrial premises with road fleets | Detergent, grease, oil, silt, grit | Vehicle washing to surface water drainage system |

*Source:* From D'Arcy *et al.* (1984).

### 3.2.2 Deepols – potable water supplies at risk

North West England provided another classic example of industrial estates as a significant water pollution problem, also in 1984, when drinking water contaminated with phenol was distributed to some two million people. The public potable water supply was the River Dee, which arises in North Wales and is abstracted upstream of Chester, NW England, for treatment and distribution there and to Merseyside and beyond. There had been a spillage of phenol at Wrexham industrial estate in North Wales which polluted a tributary stream which in turn contaminated the main river. The water treatment process included chlorination, which did nothing to reduce potential health risks of contaminants, but did give a strong readily recognised taste to the water (di-chlorophenol is the distinctive flavour of medicinal disinfectant TCP; present as well as the main constituent, trichlorophenol). That incident raised awareness of pollution sources and risks across the catchment. In particular, it highlighted the risks of industrial estate contamination of surface waters, by polluted runoff, not just licenced effluent discharges (see D'Arcy *et al.* 2000). Sadly, it was not a unique incident, and a thorough analysis of risks and implementation of continuous monitoring has produced an extensive database of contamination evidence – 'Deepols' (Figure 3.1). Episodes are classified in three categories, from high risk (1) down to low (3), and a fourth category records contamination data as precautionary evidence.

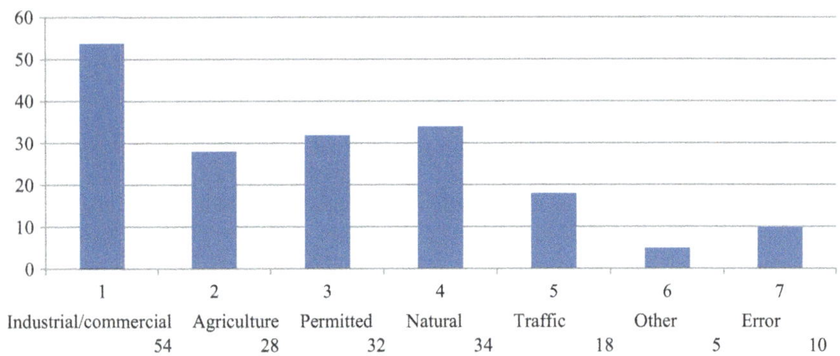


**Figure 3.1** Summary of recorded pollution incidents in the River Dee with potential to contaminate potable supplies ('Deepols'), from 1993–2010; all categories, by type of source (Environment Agency [EA], 2011).

The Deepols evidence indicates the predominance of a large industrial estate as the principal risk, with three times as many incidents (54) as, for example, traffic accidents (18), despite the Dee catchment including sections of major trunk roads into N Wales and the English Midlands. Industrial estate pollution incidents (54) exceeded the number of incidents traced to licenced effluent discharges too (32), or agriculture (28). Other categories established in the Deepols procedure are 'natural/weather related' (34) and 'other' (5) denoting incidents which were not easily categorised. The catchment is predominantly agricultural, with some uplands and forestry; it is possible the 'natural/weather related' is actually diffuse pollution. The seven types of sources are shown in Figure 3.1.

From 1993–2010, there were four serious pollution incidents (category 1, presenting a risk to potable supplies); half of them involved Wrexham Industrial Estate, and they were both involving fires. One was at a site there controlled by a waste permit. The two other category 1 incidents between 1993–2010 were for an unknown ammonia source, and a low dissolved oxygen condition in the river, cause again unknown.

## 3.3 CASE STUDY (2) DAEGU, KOREA

Sungseo Industrial Park, is a large industrial area in Daegu, the third largest city in South Korea. The industrial area is 44,457,000 m$^2$, and drains to the Nakdong River (see Figure 3.2). The number of industries there are: fabricated metal industry (1039), textile and clothing (535), transportation equipment (389), electronic electricity industry (195), primary metal (112), petrochemical industry (110), non-metal (150) and others.

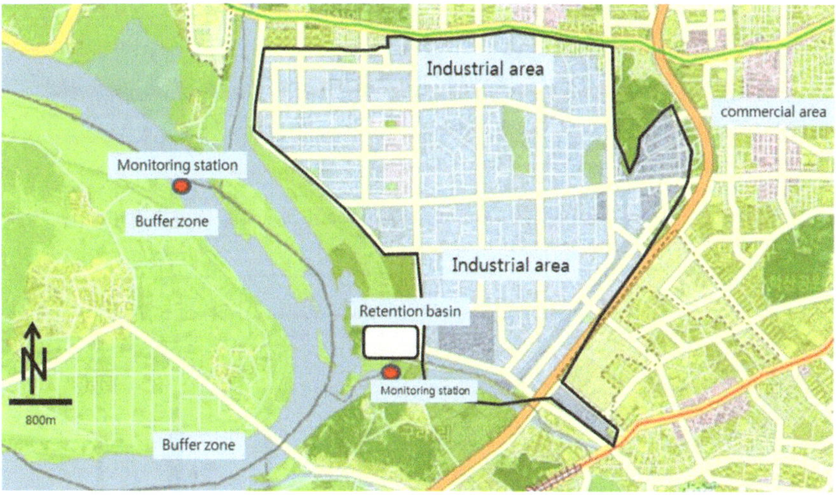


**Figure 3.2** Location of industrial areas: Sungseo Industrial Park, Daegu, Korea.

The industrial area is drained by a separate sewer system. Industrial process effluents are treated, together with sewage, in a wastewater treatment plant (WWTP), which discharges about 60,000 m$^3$/day. There is a monitoring station upstream of the industrial area, and the following figures compare quality in the Nakdong River above the industrial area, with the quality of the industrial stream which receives the industrial runoff and the WWTP effluent discharge from the district (data from Korean MOE).

In 1996 the Ministry of Environment (MOE) tightened the effluent standard to include TN and TP, contributing to water quality improvements in the watercourses. In 2006, the Korea MOE revised 'The Water Quality and Aquatic Ecosystem Protection Act' to include nonpoint source management. Accordingly, a retention basin was constructed for Daegu Industrial Park to treat the stream water contaminated by stormwater runoff. After constructing a detention basin and WWTP, the water quality of TSS, BOD and TP in industrial areas reached a similar value to that of water quality in the main stream, but TN and coliform values remained high (Figures 3.3–3.4).

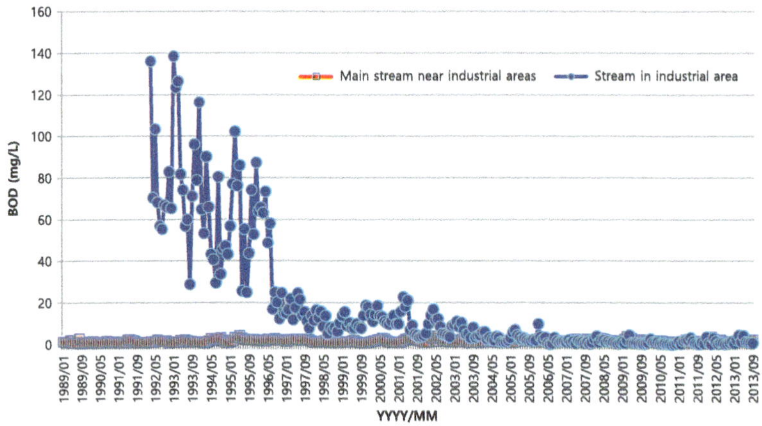


**Figure 3.3** BOD concentrations in the industrial stream draining Daegu Industrial Park, compared with the Nakdong River upstream.

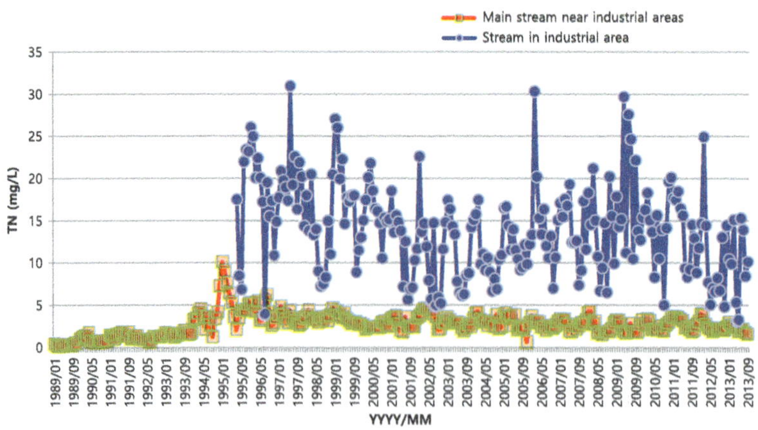


**Figure 3.4** Total nitrogen, TN, concentrations in the industrial stream draining Daegu Industrial Park, compared with the Nakdong River upstream.

Figure 3.5 shows the average yearly concentrations of water quality parameters in the Daegu industrial stream and main stream (Nakdong River) upstream, during 1989–2013 (24 years). All of the water quality parameters show higher concentrations in the industrial stream than the main stream. Metals and toxic chemicals were not detected in the main stream, but high concentrations were measured in the industrial stream. This means that even after the construction of a WWTP and retention basin in industrial areas, more still needs to be done because many toxic chemicals and metals are discharging from industrial areas by rainfall.

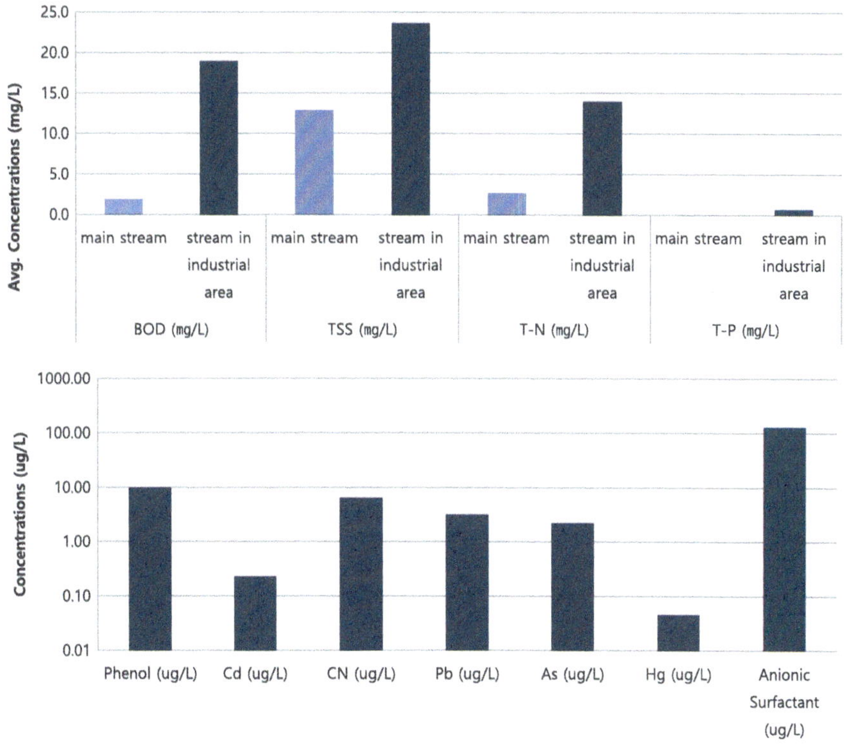


**Figure 3.5** Average concentrations of pollutants in the industrial stream at Daegu, compared with the Nakdong River upstream.

## 3.4 DISCUSSION

### 3.4.1 Oil

The most ubiquitous significant toxic pollutant in the UK case studies is oil (Table 3.5 and Wilson & Clarke, 2005). Ellis and Chatfield (2006) compared the benefits of oil interceptors and soft engineering techniques for managing diffuse sources of oil pollution. Interceptors and silt traps accumulate pollutants and require maintenance to empty them. Wetlands and swales for example, can allow the degradation of the toxic component of the contamination, the oil. The arguments developed in Ellis and Chatfield were subsequently supported by research to evaluate the relative fate of pollutants in soil/grass systems and in pond sediments and silt traps (Napier *et al.* 2009). Vegetated aerobic systems favour degradation; that should be a driver for designing SUDS infrastructure for industrial estates and premises. Degradation rates increase with temperature however, and oil contamination may be less evident in warm climates with no seasonal cold periods.

## 3.4.2 Seasonality and diffuse pollution characteristics

Monthly changes of water quality can be seen in data from the Korean MOE, comparing industrial WWTP discharges (mean values from 69 monitoring stations, urban streams (49 stations), rivers (840 stations) and lakes (189 stations), in Figures 3.6–3.7. The following water quality parameters were measured:

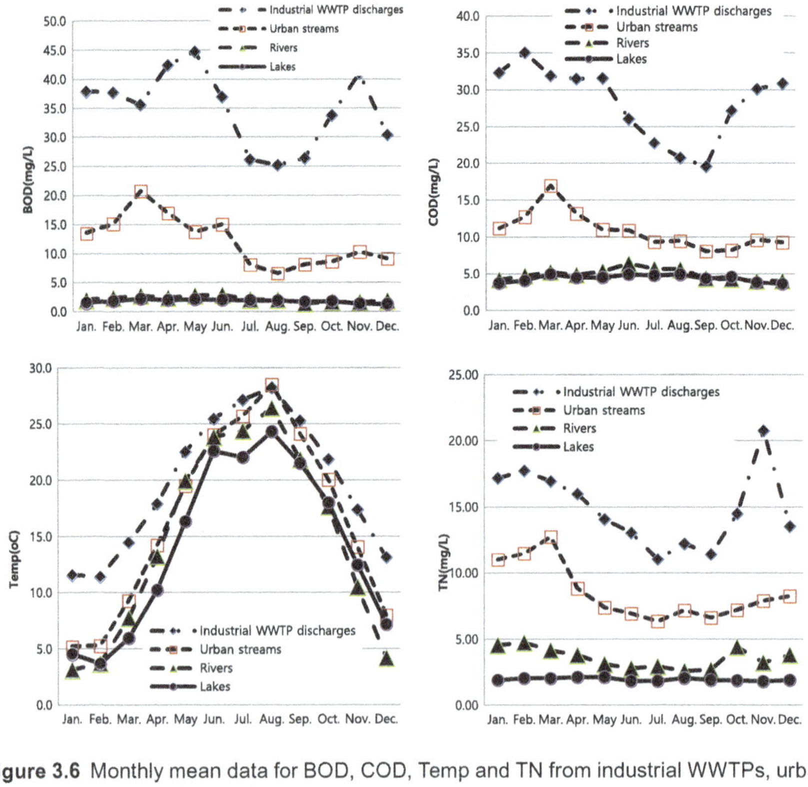


**Figure 3.6** Monthly mean data for BOD, COD, Temp and TN from industrial WWTPs, urban streams, rivers and lakes in Korea.

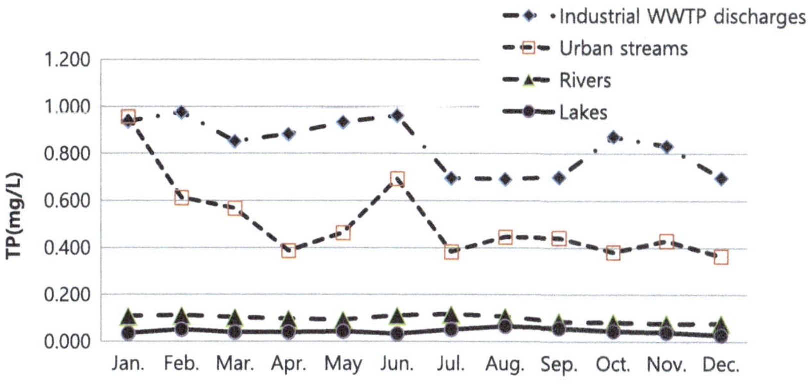


**Figure 3.7** Monthly mean data for TP from industrial WWTPs, urban streams, rivers and lakes in Korea.

pH, organics, coliforms, nutrients (2 ~ 5 times per every month), metals (4 times/year for lakes and rivers, 12 times/year for industrial WWTP discharges), other toxic chemicals (2 ~ 4 times/year).

Most pollutant concentrations in industrial WWTP and urban streams are very high compared to water quality in rivers and lakes. Even though the average concentrations during the Korean summer season (rainy periods) is lower because of the dilution effect on the point source effluent discharges, total mass loadings are very high, due to the huge amount of stormwater runoff volume. It also means that toxic chemicals and metals are discharging to the main stream during rainy periods; mobilised from the industrial areas by storm events. Limited biological data in some small watercourses impacted by diffuse pollution from industrial estates suggest the best quality is in dry summer periods, presumably associated with less frequent wash out of pollutants from the industrial areas (Gillard & Clayton, 2011).

## 3.5 CONCLUSIONS

Industrial estates are typically sources of pollution. The most numerous sources are diffuse – spread across the estate or area, and contamination is mobilised by rainfall or seepage into soil and groundwater. The worst case type of diffuse source pollution can be industrial estate hotspots. This is likely to be valid anywhere in the world, wherever a variety and significant quantity of potential pollutants are handled and processed, with outdoors loading/unloading and storage/cleaning areas, and road (or rail or boat) traffic. There are two broad categories of pollution events (albeit with great overlap grading to indistinct differences):

(a) Major accidents and incidents
(b) More or less continuous or frequent intermittent pollution; anthropogenic levels of background contamination.

If an estate is given the name 'Industrial' 'Trading' 'Business' or 'Eco-' there should be a high probability of being able to undertake reasonable business activities without coming into conflict with downstream neighbours and businesses, and without constraints on business likely to be imposed by government agencies or water utilities owing to poor design and lack of appropriate infrastructure. This is rarely the case unfortunately, unless planning requirements have specified use of SUDS infrastructure to enable site managers to see and address pollution risks on the premises under their direct control, prior to surface drainage leaving the site, and waste regulations regarding storage of materials in open vessels outdoors are in place and enforced.

## REFERENCES

Bannerman R., Baum K., Hughes P. and Graczyk D. (1983). Nationwide Urban Runoff Program. Vol. 1. Evaluation of Urban Nonpoint Source Pollution Management in Milwaukee Country, Wisconsin, Milwaukee, Wisconsin.

D'Arcy B. J. and Bayes C. D. (1995). Industrial estates: a problem. In: Proceedings of the Tenth Meeting of the Standing Conference on Stormwater Source Control, C. Pratt (ed.), School of The Built Environment, Coventry University, Coventry, ISBN 0 905949 33 1.

D'Arcy B. J., Ellis J. B., Ferrier R. C., Jenkins A. and Dils R. (2000). Diffuse Pollution Impacts – the Environmental and Economic Impacts of Diffuse Pollution. CIWEM, London, pp. 118–122, ISBN 1 870752 46 5.

D'Arcy B. J., Wilbraham D. J. and Williams B. (1984). Pollution from Industrial Estates. North West water Authority (Rivers Division), unpublished technical report RI 84/2.

Ellis J. B. and Chatfield P. (2006). The sources, pathways and management of diffuse oil pollution discharges to UK urban receiving waters. *Urban Water*, **3**(3), 165–173.

Environment Agency (2011). Freedom of Information Act enquiry. http://www.swig.org.uk/files/cms/Contamination%20Risk_2012_Skilling%20%20Warburton_1328117979.pdf.

Gillard J. and Clayton J. (2011). pers.com.

Go, S., Memon S. A. and Lee C. H. (2009). Characterization of stormwater pollutants and estimation of unit loads for road and parking lot in Gyeongan Stream Watershed. *Journal of Korean Society on Water Environment*, **25**(5), 689–696.

Kim L. H., Ko S. O., Lee B. S. and Kim S. (2006). Estimation of pollutant EMCs and loadings in highway runoff. *Korean Journal of Civil Engineers*, **26**(2B), 225–231.

Kim S. J., Shin S. M., Jeon Y. T. and Won C. H. (2012). Washoff characteristics of non-point source pollutants and estimation of unit loads in suburban industrial complex areas runoff. *Korean Journal of Environmental Impact Assessments*, **21**(2), 315–325.

Lee S. Y., Lee E., Maniquiz M. C. and Kim L. H. (2008). Determination of pollutant unit loads from various transportation land-uses. *Journal of Korean Society on Water Environment*, **24**(5), 543–549.

Mitchell G. (2005). Mapping hazard from urban non-point pollution: a screening model to support sustainable urban drainage planning. *Journal of Environmental Management*, **74**, 1–9.

MOE (1999). Guidelines of TMDL, Notification No. of 1999-143 of Ministry of Environment, Korea.

Napier F., Jefferies C., Heal K. V., Fogg P., D'Arcy B. J. and Clarke R. (2009). Evidence of traffic-related pollutant control in soil-based Sustainable Urban Drainage Systems (SUDS). *Water Science and Technology*, **60**(1), 221–230.

NFWMD (1994). St Marks and Wakulla River Resource Assessment and Greenway Protection Plan, Appendix 4. North Florida Water Management District, USA.

Novotny V. and Olem H. (1997). Water Quality: Prevention, Identification and Management if diffuse Pollution. John Wiley, New York.

Kim J.G. and Ahn, S.J. (2006). Runoff analysis based on rainfall estimation using weather radar. *Journal of the Korean Society of Civil Engineers*, **26**(1B), 7–14.

San Francisco Estuary Institute (2007). The Pulse of the Estuary: Monitoring and Managing Water Quality in the San Francisco Estuary. SFEI 532, San Francisco Estuary Institute, Oakland, CA. www.sfei.org

Wilson C., Clarke R, D'Arcy B. J., Heal K. V. and Wright P. W. (2005). Persistent pollutants urban rivers sediment survey: implications for pollution control. *Water Science and Technology*, **51**(3–4), 217–224.

# Chapter 4

# Risk assessments – trader activities and water pollution

*Zorica Todorovic* and Arthur Thornton*

*Atkins Limited, UK*

**Corresponding author: zorica.todorovic@atkinsglobal.com*

## 4.1 INTRODUCTION

The European Water Framework Directive (WFD) requires that all water bodies need to achieve 'good' status by 2027. In order to meet this requirement, sources of pollution to the water bodies have to be identified and adequate mitigation measures put in place so that the relevant water quality standards are met.

Surface water from industrial estates has the potential to be a major source of diffuse pollution in certain urban areas. The approach to mitigate pollution from these sites needs to be underpinned with robust evidence based on a good understanding of the sources/types of pollution and the pathways of transfer in order to devise comprehensive management measures.

This paper presents the complexity of different sources of pollution that need to be taken into account when assessing both the impact of an industrial estate on the receiving watercourses and the measures to be taken to mitigate pollution based on the identified associated risks.

This approach has been tested at a number of industrial estates in Scotland that discharge, via surface water outfalls, directly into watercourses. The main driver for implementing mitigation measures was the WFD that has been transposed into the Water Environment (Controlled Activities) (Scotland) Regulations

© IWA Publishing 2017. Wealth Creation without Pollution: Designing for Industry, Ecobusiness Parks and Industrial Estates
Brian D'Arcy, Lee-Hyung Kim, Marla Maniquiz-Redillas
doi: 10.2166/ 9781780408330_53

2005 – more commonly referred to as the Controlled Activities Regulations or CAR. The Scottish Environment Protection Agency (SEPA) has been appointed as a competent authority responsible for enforcing the CAR in Scotland. Under this legislation, SEPA exercised its discretionary power by identifying industrial sites where discharges from surface water outfalls are considered to potentially contain pollutants from the list of 'dangerous substances', which could be detrimental to the water quality of the receiving watercourses.

The aim of the study was to determine the appropriate level of control of surface water quality from trader estates that would be appropriate to the relative threat that the trader estates posed to the receiving water. The project deliverables were site specific Surface Water Action Plans.

The overall approach in developing Surface Water Action Plans was to identify:

- Trader activity risk assessment;
- Identification of abnormal or potentially higher risk site activities;
- Risk minimisation plans to normalise the risks at the specific sites;
- Derived management responses for the identified residual risks that remained;
- Evaluation of the relative contribution of the trader estate compared to other sources.

This paper sets out the detailed approach and presents a comprehensive application of risk management interventions that demonstrate a collaborative and holistic approach to surface water management arising from industrial estates.

It is recognised that sustainable drainage systems (SUDS) may, in some instances, be part of the solution. It is critical that the disciplined approach outlined is followed to facilitate the selection and design of the most appropriate SUDS options.

## 4.2 TRADER ACTIVITIES AND POLLUTION RISKS

### 4.2.1 Activities and pollution risks – desktop study

The first step in assessing pollution risk is to undertake assessment of the risks based on the activities at the industrial estate. It identifies traders' activities at industrial estates and the likely magnitude of their impact on surface water quality. Key industrial estate activities investigated in detail in this assessment are presented in Table 4.1.

Unless there are specific recorded incidents at the industrial estate suggesting pollution from a specific source, focus should be directed to an assessment of key pollutants that are most likely to be present in surface water runoff at a typical industrial estate (Gromaire *et al.* 2001; Davis *et al.* 2006; Rule *et al.* 2006; EA, 2007; Stein *et al.* 2007; Napier *et al.* 2008; Park & Stenstrom, 2008; McKenzie *et al.* 2009; Marshales & Viklander, 2011). These are summarised in Table 4.2.

**Table 4.1** Key industrial activities.

| Industrial Estate Activities | Key Investigations on Site |
|---|---|
| Vehicle movement | Assess traffic density across the site<br>Identify traders associated with distributing goods, depots and courier-type services<br>Assess frequency of road sweeping |
| Chemical storage | Identify chemical storage areas<br>Investigate storage procedures and type of chemicals stored |
| Vehicle wash | Identify areas of vehicle washing<br>Assess frequency of washes per trader<br>Identify if a surface water collection system is in place<br>Identify type of detergents used |
| Material storage | Identify material storage areas<br>Investigate storage procedures and type of materials stored |
| Operational spills risk | Identify manufacturing sites<br>Identify operations related to re-fuelling and pumping of liquids |
| Wastewater misconnections | Investigate history of sewer misconnections<br>Inspect manholes visually |

### 4.2.2 Prioritisation of risk areas

Prioritisation of the risk areas has been done based on key activities. Based on literature (Davies *et al.* 2001), a typical (light) industrial estate is a self-contained series of units predominantly non-retail on the perimeter of towns and cities.

The risk of surface water contamination arising from trader activities was prioritised in accordance with the following hierarchy:

- *High risk (red)*: contamination is very likely to occur and requires further investigation on site (possibly including sampling) and further research on substances released via surface water sewer. These areas are usually related to inadequate site management practices on site which affect areas of chemical and material storage and areas of operational spill risks. Key hot spot areas on industrial estates are storage and service yards, historic contaminated land and any areas with illegal waste disposal and general industrial litter. Chemical/oil spills that can be the result of various industrial activities, for instance dispensing oil, antifreeze, and other potentially hazardous liquids, can result in spills and leaks around the dispensing area. Another area focused on was leakage or spillage occurring around tanks from filling, dispensing, and deterioration of pipe connections or failure of secondary containment. Sewer misconnections are considered high risk of contamination with raw sewage and the possibility of blockages in the surface water drainage network and related risk of localised flooding.

**Table 4.2** Key surface water pollutants and their sources.

| Pollutant/ Indicator | Sources at Industrial Estates |
|---|---|
| BOD and COD | Organic matter disposed via waste disposal and recycling centres<br>Possible release of toxic leachate via surface water drainage<br>Organic matter associated with solids washed off impermeable surface areas during a rain event<br>Corrosion Inhibitors used at manufacturing sites |
| TSS (Total Suspended Solids) | Road runoff<br>Runoff from impermeable surfaces, including roads, car parks, service & storage yards<br>Car washing & maintenance<br>Equipment or vehicles stored outside<br>Vehicle repair/brake shoe replacement<br>Material storage: stripping metal or wood surfaces outdoors<br>Atmospheric deposition<br>Storage of general waste or food waste outside in dumpsters |
| Total Heavy Metals, mainly zinc, copper, lead | As above for sources of TSS as most metals are associated with solids<br>Exposed copper/galvanised piping, galvanised building sidings/ roofing, or exposed copper, brass, or zinc coated materials exposed to stormwater<br>Car washing & maintenance<br>Vehicle repair/brake shoe replacement<br>Material storage: stripping metal or wood surfaces outdoors<br>Replacement or storage of lead/acid or nickel/cadmium batteries or long time storage of vehicles or powered equipment outside |
| PAHs (Polycyclic Aromatic Hydrocarbons) | Equipment or vehicles stored outside<br>Car washing (two sources: (i) cleaners containing mineral spirits/oil or petroleum products, (ii) washing off lubricating oils and dust)<br>Vehicle maintenance, equipment maintenance, involving grease<br>Parts & equipment cleaning outdoors<br>Oil (& other fluids) dispensing & outside storage<br>Wrecked or damaged vehicle storage<br>Pumping liquids from storage tanks into site buildings or into vehicles |
| Ammonia | Storage of general waste or food waste outside in dumpsters<br>Organic matter disposed via waste disposal and recycling centres<br>Possible release of toxic leachate via surface water drainage |

- *Medium risk (amber)*: contamination is likely to occur. Traders' operations and other site activities identified as having medium risk need to be checked on site and further reference literature data need to be gathered to assess the requirement for treatment by SUDS. These areas include mainly car parks and roads. Key activities that bring attention to possible surface water pollution are *ad hoc* uncontrolled car washing, vehicle movement and impermeable surface runoff during a storm event.

- *Low risk (green)*: contamination is not likely to cause concern; no further investigation is required. Usually these areas are associated with roofs and building sidings (depending on the type of materials and age, these could be considered as medium risk) and landscape areas. Site management activities such as cleaning of road sides, gully pots, existing interceptors, regular removal of litter, preservation of landscape areas and general aesthetics of the site area are all indicators of how well the site is managed. Additionally, storage and service yards of well-managed traders that can give rise to some residual diffuse pollution are considered low-risk areas.

## 4.2.3 Verification through site visits

The desktop analyses were verified through site visits. The site visits focused generally on high risk areas and the locations with 'atypical' characteristics were identified (Figure 4.1).

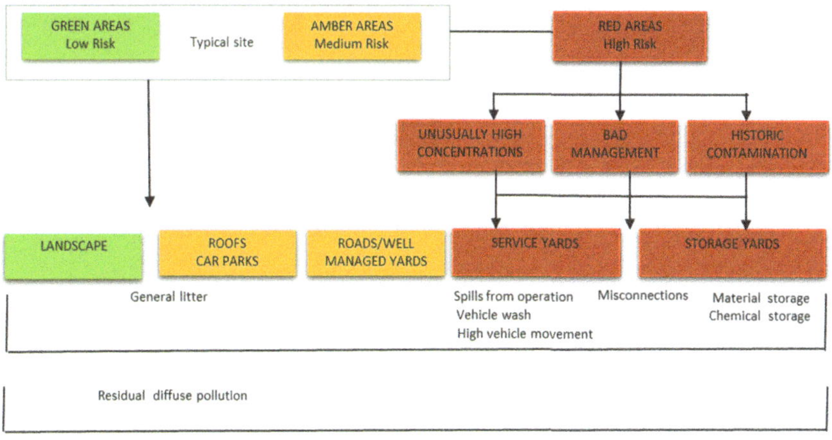


**Figure 4.1** Pollution risk areas and activities.

Maps of the industrial estates (Figure 4.2) were produced for each site marking clearly the areas based on risk of pollution. These areas would require additional pollution control measures to reduce their 'a typicality' and normalise the sites into the typical (light) industrial estates category. Very often the site visits revealed high risk areas in places that were not previously identified by the desk-top study. Figure 4.2, represents a typical situation, but is for illustrative purpose only and does not identify the actual trader risks or contributions.

**Figure 4.2** Pollution risk areas across the industrial estate.

## 4.3 NORMALISATION OF RISKS

Identified high risk areas need to be targeted by Site Specific Measures in order to 'normalise' the industrial estate (Figure 4.3).

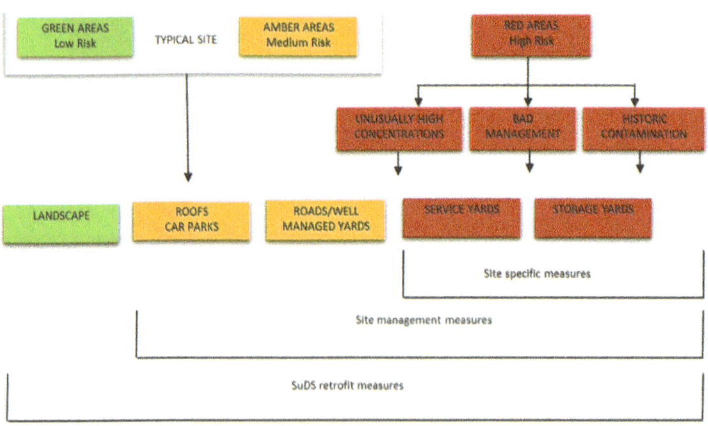


**Figure 4.3** Mitigation measures based on pollution risk.

For example, the non-compliant traders need to change their practices through raising awareness, education and regular auditing.

Open storage yards with potential of leakage should be roofed. High risk areas prone to spillages should introduce measures in the form of containment sumps and berms, bins with screens, oil interceptors, compost filters, settling tanks, etc. A good understanding of the drainage system is required to assess pathways of pollution and provide adequate interception and containment. Catchpits should be provided in locations where increased sediment load is suspected. As a last resort, diversion of the first flush of surface water from these areas to the foul drainage system could be considered, providing adequate capacity is available.

Potential misconnections or leaks of historic contaminated areas via groundwater that were identified at the outfalls outside of rainfall events need to be dealt with separately.

## 4.4 DERIVED SITE MANAGEMENT RESPONSES FOR THE IDENTIFIED RESIDUAL RISKS

Once the sites were 'normalised', a set of specific measures were considered.

### 4.4.1 Surface Water Action Plan

Site Management Measures were outlined in Surface Water Action Plans (SWAPs) that were put in place for each industrial estate. Site management includes solutions that do not involve construction of a feature but are changes in the way the site is managed. In order to minimise pollution from the site, there was a need to immediately establish a good site management programme which included:

- Regular road sweepers, gully pot emptying and litter cleaning are recommended for the whole site, especially at the road sides;
- Regular inspection as a way to ensure that the best management practice is maintained;
- Up-to-date guidance on good housekeeping delivered to all traders;
- Collaborative work with external organisations to appropriately manage their assets which impact on water quality on site (i.e., road drainage, garbage bin washing, car washing);
- Once site measures are in place, clear guidance for new connections and alterations of existing plots and good practice guidance for redevelopments to be developed and issued to ensure best management practice is applied for redeveloped plots and adequate mitigation measures are applied at source from the onset of new businesses on site;
- Ensuring best practice through planning approvals;
- Development of site-wide emergency plans, which are aimed to prevent major pollution spills during site-wide incidents like floods or fires;
- Development of standards or code of practice for car washing at industrial estates, which will significantly reduce pollution load at the estates visited.

It has been recognised that different actions within the SWAP would rest with different stakeholders. To be effective, all stakeholders will need to work together and fully utilise the powers and mechanisms which are available to them.

### 4.4.2 SUDS retrofit measures

Once the site management measures are in place, additional SUDS measures could be proposed to mitigate chronic background pollution. Types of preferred SUDS for a particular site would depend on space availability and could be targeted to remove specific types of pollutants from specific areas (Barret *et al.* 1998; Bäckström, 2003; Escarameria *et al.* 2006; Todorovic *et al.* 2008, 2010; Bacci *et al.* 2010; D'Arcy *et al.* 2010). Sometimes, further sampling of water quality might be required to facilitate proper source apportionment. A typical SUDS solution for industrial estates includes:

- On site source control SUDS measures such as: green roofs, soakaways, filter strips, trenches, swales, bioretention areas, reed-beds or porous pavements; or
- Installation of end of pipe SUDS by diverting surface water sewers close to the outfall into the retention and detention ponds or basins to agreed standards.

Incorporating sustainable drainage system (SUDS) to treat surface water runoff (e.g. Figure 4.4) at an industrial estate is recommended only for low to medium risk areas of pollution, when the site is already 'normalised'. That means that they could assist in mitigating pollution from a typical light industrial estate, which is well managed and has no historical problems with contamination. Lack of or insufficient pollution mitigation measures for high risk areas could jeopardise the efficiency of the SUDS measures applied for the site in general.

**Figure 4.4** Examples of SUDS at industrial estates.

## 4.5 RELATIVE CONTRIBUTION TO WIDER CATCHMENT POLLUTION

Which level of pollution control measures is required was based on the relative contribution of each industrial estate to the wider catchment. The study considered how significant the discharge from the industrial estate was by determining its overall contribution to the WFD monitoring point downstream of the surface water discharge. The comparative loads of pollutants arising from the estate were calculated using literature values for a light industrial estate (Rule *et al.* 2006) so as to determine if it were likely to result in failures or contribute to failure at the WFD monitoring point. The methodology adopted was to:

- Determine the geographical area of the industrial estate and assess the impermeable area to determine the run-off area: This was achieved by using satellite map data and drainage plans.
- Determine the hydrological definition of the catchment geographical area of the WFD monitoring point and flows at the point: LowFlows V1.0.6 was used to define the catchment boundaries upstream of the discharge. This is a programme developed by Wallingford Hydrosolutions (UK) to estimate flows at specified locations on a watercourse. A second programme was also used to estimate flows called Flood Estimation Handbook (FEH) (Centre for Ecology and Hydrology [CEH], 1999) which has been designed for the purpose of estimating high flows for flooding studies. These programmes use different digital terrain models (DTMs) to define the catchment draining to a specified location on a watercourse. The catchment boundary obtained using LowFlows was validated by comparing it with FEH. These were found to be in agreement (±10%) for all but the localised catchment.
- Consider the % urbanisation, including likely contributions such as motorways, railway lines and potential contaminated land: This was achieved by overlaying the boundary from the hydrological map determined above on a satellite map of the urban area, including significant contributions from highways, rail or significant other industrial areas.
- Determine the chemical and ecological status at both the WFD monitoring point and the trader discharge point. Water quality data were available from the Scottish Environment Protection Agency for the pollutants identified in the literature. River Basin Management Plans (RBMPs) prepared as part of the WFD look not only at the chemical status of surface water bodies but also their ecological status. The ecological status is an integrated assessment which includes biology, chemistry (physico-chemical and specific pollutants), hydrology and morphology. The classification upstream of the discharge and downstream were considered.

There are instances where the size of the estate compared to the whole WFD catchment and the intense urbanisation of the overall catchment will clearly indicate that the contribution of the estate, in which the high risk activities have been identified and controlled, will be of very little significance to the WFD monitoring point. However, the impact on the local receiving water may need to be considered, especially for the impact of fuel oil that can substantially visually impact receiving water.

In most cases the initial assessment of the site indicated that post 'normalisation' the sites would be considered no more impactful than a typical urban area so the selection matrix for SUDS solutions that are conventionally used for other urban areas were deemed appropriate.

## 4.6 CONCLUSIONS

Although the nature of the economic activity of the traders was related to the risk of diffuse pollution on site, the management of processes involved have a much bigger impact on the surface water quality.

There are some ubiquitous activities, such as vehicle washing, that are largely uncontrolled and potentially locally impactful.

Having 'normalised' the quality through the improved risk identification and control, the surface water runoff is similar to that anticipated from other urban areas.

Having in mind sizes of the industrial estates in comparison to the area of the wider catchments, the contribution of pollution to the WFD waterbody in most cases was relatively low, though local streams may need higher levels of protection, especially from the hydrocarbons such as oil, as small spills may have a disproportionate visual impact.

However, accidental spillage risks need to be managed through site specific measures (for example bunded areas for chemical holding, etc.) and need site specific review and if necessary monitoring of water quality to design site specific measures.

The SUDS technologies that are commonly used within the domestic space are effective options after risk normalisation at industrial estates and could improve water quality that is discharged into the watercourses.

The collaboration of multiple stakeholders in a Surface Water Action Plan is critical and can drive improvement through behaviour and site management practice (e.g., storage areas, gully pot cleaning etc.).

## REFERENCES


Bacci F., O'Reilly S., Morgan G., Heffron J., O'Halloran J. and Foley A. M. (2010). Abatement Technologies for Road Surface Run-off. Proceedings of ITRN 2010, Irish Transport Research Network, University College Dublin (UCD), Dublin.

Bäckström M. (2003). Grassed swales for stormwater pollution control during rain and snowmelt. *Water Science and Technology*, **48**(9), 123–134.

Barrett M. E., Walsh P. M., Malina, Jr., J. F. and Charbeneau R. J. (1998). Performance of vegetative controls for treating highway runoff. *Journal of Environmental Engineering, ASCE*, **124**, 1121–1128.

CEH (1999). Flood Estimation Handbook. Centre for Ecology and Hydrology, Wallingford, ISBN 9781906698003.

D'Arcy B., Todorovic Z., Shaffer P., Hemingway A. and Martin R. (2010). Trading Estates Best Practice. IWA DIPCON 2011, New Zealand.

Davis A. P., Shokouhian M. and Ni, S. B. (2001). Loading estimates of lead, copper, cadmium, and zinc in urban runoff from specific sources. *Chemosphere*, **44**, 997–1009.

EA. (2007). The unseen threat to water quality Diffuse water pollution in England and Wales report. http://www.environment-agency.gov.uk/research/library/publications/33983.aspx

Escarameia M., Todd A. J. and Watts G. R. A. (2006). Pollutant Removal Ability of Grassed Surface Water Channels and Swales: Literature Review and Identification of Potential Monitoring Sites. Highways Agency published report, UK.

Gromaire M. C., Garnaud S., Saad M. and Chebbo G. (2001). Contribution of different sources to the pollution of wet weather flows in combined sewers. *Water Resources*, **35**(2), 521–533.

Marsalek J. and Viklander M. (2011). Controlling contaminants in urban stormwater: linking environmental science and policy. *On the Water Front*, J. Lundqvist (ed.), Vol. 2, Stockholm International Water Institute, Stockholm, Sweden, pp. 100–107.

McKenzie E. R., Money J. E., Green P. G. and Young T. M. (2009). Metals associated with stormwater-relevant brake and tire samples. *Science of the Total Environment*, **407**, 5855–5860.

Napier F., D'Arcy B. and Jefferies C. (2008). A review of vehicle related metals and polycyclic aromatic hydrocarbons in the UK environment. *Desalination*, **226**(1–3), 143–150.

Park M-H. and Stenstrom M. (2008). Comparison of pollutant loading estimation using different land uses and stormwater characteristics in Ballona Creek Watershed. *Water, Science and Technology*, **57**(9), 1349–1354.

Rule K. L., Comber S. D. W., Ross D., Thornton A., Makropoulos C. K. and Rautiu R. (2006). Diffuse sources of heavy metals entering a wastewater catchment. *Chemosphere*, **63**, 64–72.

Stein E. D., Tiefenthaler L. and Schiff K. (2007). Land-Use-based Sources of Pollutants in Urban Stormwater. Proceedings of 80th Water Environment Federation Technical Exhibition and Conference, San Diego, California.

Todorovic Z., Barbarito B., D'Arcy B. and McLean N. (2010). SUDS Retrofit in Scotland to Treat Diffuse Pollution from the Industrial Estates. DIPCON 2010 Conference, Quebec.

Todorovic Z., Reed J. and Taylor L. (2008). SUDS Retrofit for Surface Water Outfalls from Industrial Estates: Scotland Case Study. 11th International Conference on Urban Drainage, Edinburgh, Scotland, UK.

# Chapter 5

# Green industry concept and practices

*Lee-Hyung Kim and Marla C. Maniquiz-Redillas*

*Department of Civil and Environmental Engineering, Kongju National University, Cheonan, Chungnamdo, Korea (E-mail: leehyung@kongju.ac.kr, marlcmani@kongju.ac.kr)*

## 5.1 SUSTAINABLE DEVELOPMENT GOALS AND SUSTAINABLE INDUSTRY

Urbanization and industrialization are inevitable processes that enhance the quality of life. Urbanization is a process in which the natural cover is transformed into an artificial one with the aim of providing a foundation for human activities and living spaces. Unfortunately, in the course of the urbanization process, the environment becomes polluted and the ecosystem is damaged. Highly impermeable surfaces, such as roads, buildings, parking lots, etc., in urban areas cause a wide array of problems, including alteration of the natural water circulation, increased pollutant discharge, increased urban flooding, lowered groundwater level, heat island effect, and habitat destruction resulting from reduced green spaces. Meanwhile, industrialization is a process in which the goods and services required in our lives are produced and provided; this process poses an even greater threat to the environment and to our ecosystem than urbanization does.

Today, many developing countries are continuously pursuing industrialization as a way to overcome poverty and increase the standard of living so as to achieve a decent and quality way of life, while the developed countries are also expanding their industrialized platforms to further increase the quality of life of their people. Due to

© IWA Publishing 2017. Wealth Creation without Pollution: Designing for Industry, Ecobusiness Parks and Industrial Estates
Brian D'Arcy, Lee-Hyung Kim, Marla Maniquiz-Redillas
doi: 10.2166/9781780408330_65

the ever-growing thirst for industrialization in the world, however, many countries are faced with serious environmental degradation and resource depletion issues. In an effort to solve such problems, United Nations (UN) members started discussions on Sustainable Development Goals (SDGs) in Rio +20, which was held in Brazil in 2012; the SDGs were finally adopted by the UN General Assembly in September 2015. The SDGs are a way of development that suits the current generation's needs while not harming the potential to meet the needs of the next generation.

'Social development', 'Economic growth', and 'Environmental protection' are the key elements of SDGs. Currently, there are 17 goals and 169 targets set during the period of 2016–2030 (UN, 2016). Most of the SDGs are related to *industry*, *economy*, and the *environment* and in order to achieve the three key elements, green industry must be established. The purpose of SDGs is to both achieve environmental protection and economic growth at the same time which can be described as *sustainable industry*. Many businesses are now shifting from their conventional ways, focused on resource consumption and maximization of production, to a new paradigm in which production, environmental protection, and circulation of resources considering the limitations of nature can coexist. The idea of sustainable industry is important in that it can provide a comprehensive, industry-wide approach to environmental problems as well as helping individual businesses tackle environmental issues on their own. In other words, many individual companies can apply the lessons they have learned from the system of circulation in nature to the building of an interconnected production system under the framework of sustainable industry. This perspective, which aims to establish an industrial complex that has characteristics similar to those of an ecosystem, is now being vigorously promoted by both the developed and developing countries, along with the SDGs.

## 5.2 GREEN INDUSTRY CONCEPT

Green industry means a type of industry based on an eco-friendly system in which the existing industrial structure, kept intact, is restructured into a more environmentally-friendly system. Green industry pursues sustainable production and consumption patterns, considering elements such as resource and energy efficiency, low carbon and waste emission, and non-polluting and safe industry through lifecycle management. The green industry agenda includes the 'greening' of industry meaning resource production and environmental efficiency for all types of industries which is enhanced on a continuous basis. In addition, green industry generates environment-related goods and services, such as waste management and recycling services, renewable energy technologies, and environmental analysis and advisory services (United Nations Industrial Development Organization [UNIDO], 2011).

Green industry guarantees a consistent improvement in environmental performance in all industries, regardless of sector, size, or location. It includes efficient use of resources in processes and production, phased reduction in toxic substance discharge, replacement of fossil fuels with renewable energy sources, enhanced occupational health and safety, and commitment to and action toward fewer environmental impacts, along with increased liability of manufacturers and overall risk reduction efforts (Figure 5.1).

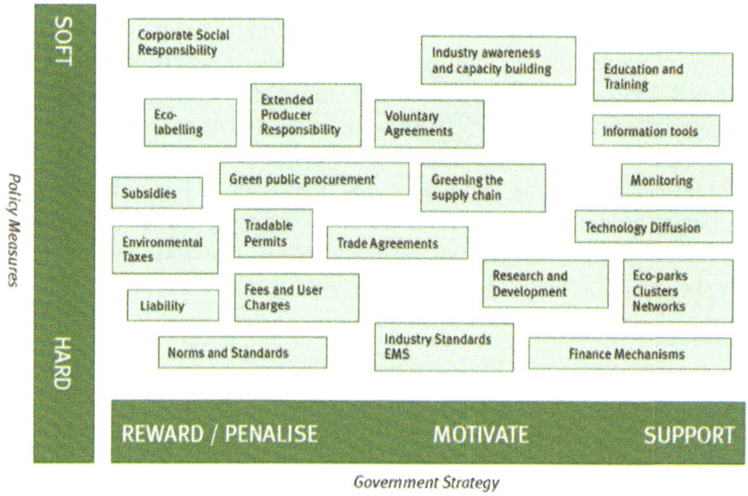


**Figure 5.1** Policy Matrix for Green Industry (UNEP/CSCP, 2007; UNIDO, 2011).

In order to achieve sustainable performance in the green industries, systematic and comprehensive methods must be adopted; these methods will also serve as the foundation for a new business model. Efforts to establish a closed-loop circular production system look to the recycling of disposed products and to new resources for production (Figure 5.2).

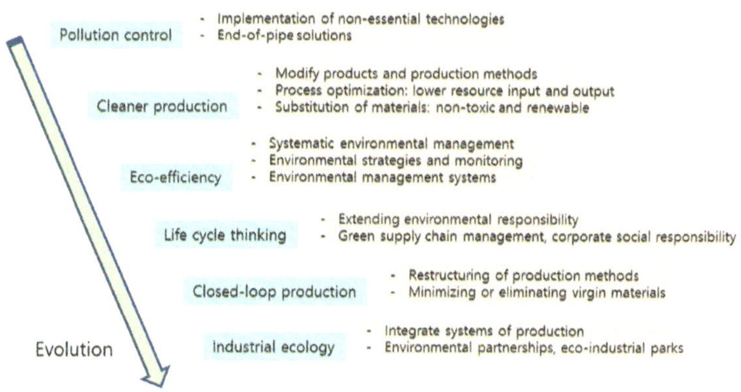


**Figure 5.2** Evolution of sustainable manufacturing concepts and practices (Organisation for Economic Co-operation and Development [OECD], 2010).

## 5.3 GREEN INDUSTRY PRACTICES

The ecosystem consists of biotics and the environment, and is run by mass circulation and energy flow. An industrial complex has mechanisms similar to that of the ecosystem, since it is a venue for business activities in which goods are produced using raw materials; this system resembles the mass circulation and energy flow of nature. Based on this perspective, the notion of industrial ecology is introduced. Industrial ecology is a field of study systematically exploring the use and flow of substances and energy in manufacturing and its processes, in industries, and in the economy as a whole at local, regional, and global levels (Gibbs & Deutz, 2005). Industrial ecology is focused on potential roles of businesses in reducing environmental pollutant loads, which are created during the lifecycle of product manufacture, and which range from the export of raw materials and manufactured goods to the use of products and on to waste management (Figure 5.3).

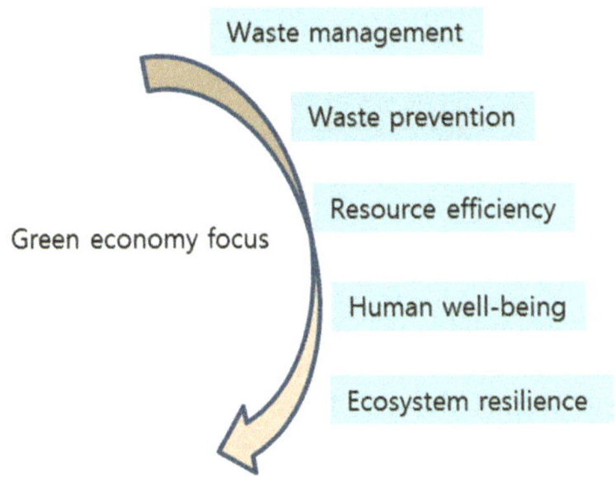


**Figure 5.3** Green economy as an integrated framework for policies on material use (European Environment Agency [EEA], 2016).

### 5.3.1 Environmental accounting (EA)

Environmental accounting (EA) is a useful tool that can help in the decision-making process of private sectors. Whether a business seeks pollution prevention or sustainability of its operation, EA can help in the realization of financially feasible environmental innovations. Environmental management accounting (EMA) is a concrete application of environmental management for identification, analysis, and use of key physical and monetary information during the internal decision-making process. Physical information includes information on the use and flow of energy, water, and materials, and also on the final products; monetary information refers to information on costs, revenues, and cost-reduction methods regarding the environment (Jasch, 2001). Material flow cost accounting (MFCA) is one of the EMA methods designated by the ISO standard; it focuses on the discharge of waste and pollutants by businesses and the tracking of their non-products. This method can contribute to reducing the use of materials, energy, and water while improving the productive use of materials, thereby helping to improve the economic and environmental effects of a nation (Figure 5.4). In a country with a relatively low labor cost, use and loss of materials and energy is a vital factor in cost-effectiveness improvement. Thus, MFCA is an important tool to increase resource efficiency in developing and emerging economies (Kokubu *et al.* 2009).

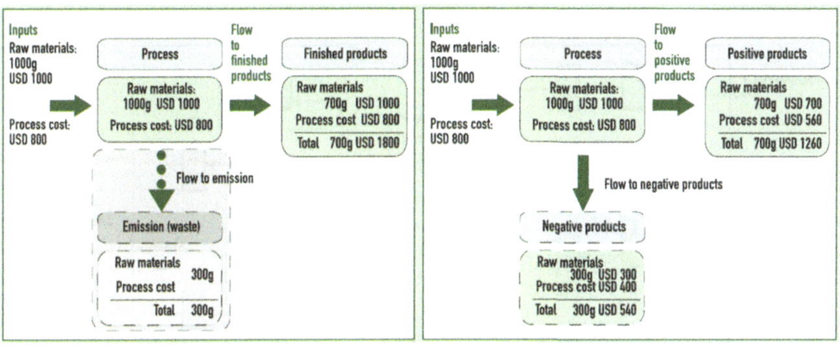


**Figure 5.4** Comparative example of conventional cost accounting (*above left*) and material flow cost accounting (*above right*) (Kokubu *et al.* 2009).

## 5.3.2 Eco-industrial park (EIP)

To develop industrial parks that use a huge quantity of energy and industrial water and also generate waste by-products into sustainable industrial parks, it is necessary to form an eco-friendly eco-industrial park (EIP). EIPs are industrial parks designed, developed and operated for a cleaner production by means of adopting the concept of industrial ecology. The EIP was first presented in the United Nations Conference on Environment and Development at Rio de Janeiro, Brazil in 1992. The universal concept of EIP used internationally is the concept first proposed by the Indigo Development Team in 1992 and this concept was expanded by the United States Environmental Protection Agency (US EPA) in 1995. The concept of EIP was improved as depicted by Ernest Lowe in the Eco Industrial Handbook published by the Asian Development Bank in 2001.

### *5.3.2.1 Planning the new industrial park*

The EIP is an industrial complex composed of the mechanisms of the ecosystem that is based on organic relations that minimize the use of materials and energy, and the amount of pollutants generated via ecological interconnectedness among businesses. In such a complex, more efficient environmental improvement activities and increase in productivity can be expected when compared to a single business attempting to do such jobs alone. The support that a nation can provide to such a complex activity can bring more achievements than any help given to a single company. The most distinctive difference between the conventional industrial complex and the EIP is that while the network formed in the conventional industrial complex focuses on raw materials and products, businesses in the EIP forge a network based on by-products and wastes (Figure 5.5). The EIP can be called an industrial symbiosis, virtual EIP, or eco-industrial network depending on the content or the type of connectedness forged among businesses (Leuenberger, 2015).

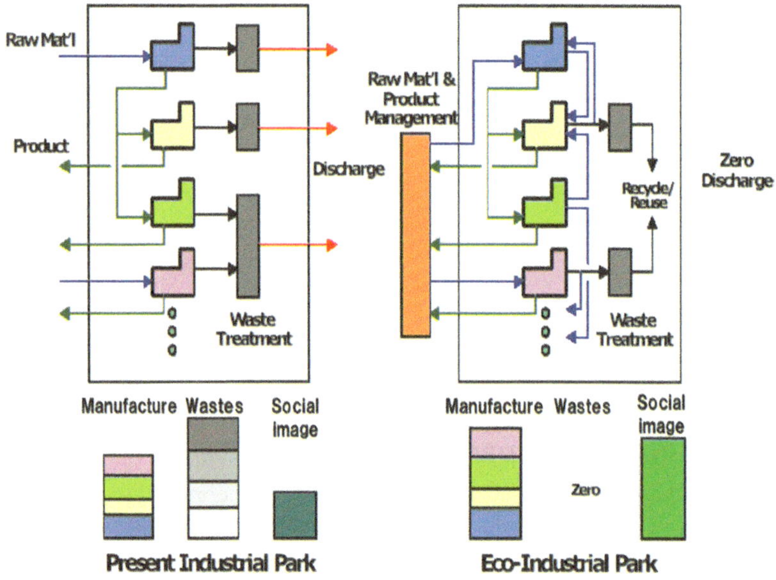


**Figure 5.5** Schematic diagram of eco-industrial park (EIP, 2005).

A comparison of the characteristics of conventional industrial parks and EIPs is shown in Table 5.1. In general, parks having only recycling and environmental technology-related firms, green product manufacturers, and a single environmental theme (e.g., parks with firms using solar energy) are not classified as EIPs. The most important aspect in building an EIP is that firms voluntarily participate in network formation, and the industrial park infrastructure is governed by low impact development (LID) that seeks to minimize impact on the environment. To this end, it is necessary to guarantee the mutual interests of the participating firms.

### *5.3.2.2 Transforming existing industrial parks*

For the industrial parks that are already created, it is important to form an eco-industrial network among firms regarding matters, energy, and by-products.

Generally, in parks where firms in the same industry are gathered together, it is impossible to exchange by-products and is thus, not easy to build an eco-industrial network. The method of transforming existing industrial parks to EIPs is shown in Figure 5.6. The transformation process of existing industrial areas/parks/clusters into EIPs involves various aspects such as infrastructure development, planning, estate management and capacity building. To efficiently transform existing industrial parks to EIPs, it is necessary to apply green infrastructure. To this end, activities shown in Table 5.2 are needed.

**Table 5.1** Characteristics of conventional industrial park and eco-industrial park (EIP).

| Parameter | Conventional Industrial Park | Eco-industrial Park (EIP) |
|---|---|---|
| Selection criteria | Economic feasibility | Economic and environmental feasibility |
| Connectivity | Raw materials or products | Raw materials, products, by-products, wastes, etc. |
| Management of wastes | Individual or joint treatment | Reuse as resources |
| Management authority | Industrial complex corporation | Industrial complex corporation and independent organization |
| Waste generation | High | Low |
| Relationship with local communities | Many civil complaints | Participation of local communities |
| Social image | Pollution sources | Environmental and social harmony |
| Components | Request for quotation | Spontaneous participation |
| Storm drainage | Fast drainage by storm pipe | Natural water circulation system |
| Soil cover | Highly impermeable pavement | Highly permeable pavement |
| Public spaces | Low | High |
| LID application | Low | High |

## 5.3.3 Stormwater management

The artificial environment, such as the industrial complex, causes serious air, water, and ground pollution due to the various types of wastes and pollutants it generates. In order to reduce the various environmental impacts, ecological engineering is being introduced to the industrial complex. The concept of ecological engineering was first introduced in 1963 by Howard Odum, an American ecologist. At the initial stage, ecological engineering suggests natural energy as a controlling factor to manage and control the environmental system

(Odum & Odum, 2003). The recent concept of ecological engineering is more about designing a sustainable ecosystem in which human society is integrated with nature so that benefits are brought to both humanity and the natural environment. The final goal of ecological engineering is to restore the ecosystem, long-disturbed by human activity, as well as to develop a sustainable eco-system in which human and ecological values go hand-in-hand. In the 21st century, landscaping, civil engineering, environmental, design, and energy technology are grafted with ecological engineering in a wide range of industrial complex and other social infrastructure development projects.

**Figure 5.6** Transforming existing industrial parks to eco-industrial parks (EIPs) (CETP: common effluent treatment plants; GI: green infrastructure; GIS: geographic information system).

Recently, in some developed countries, including the United States, the adoption of ecological engineering onto the social infrastructure sector in such areas as landscaping, civil engineering, and design, and energy technology, instead of the conventional plan-based approach used in land-use planning, has been actively promoted. This is called 'green infrastructure', a concept in opposition to gray infrastructure, which is based on the conventional social infrastructure development methods that do not take into account ecology or the environment.

**Table 5.2** Relevant activities for transforming existing industrial park to eco-industrial park (EIP).

| Parameter | Content |
|---|---|
| Setting up of common effluent treatment plants (CETPs) | • Includes sewerage system, disposal system, recycle/reuse – Activities include undertaking feasibility studies<br>• Prepare the project development module with relevant business model for setting up of CETP<br>• Implement the optimal technology |
| Stormwater drainage system | • Assess and restore the natural water circulation system by increasing infiltration capacity and green spaces<br>• Implement the possible green infrastructure technology |
| Plantation/green belts | • Identify the potential areas (block plantation, avenue plantation) by environment management cells (EMCs) at industrial parks<br>• Implement the possible plantation/green belts technology |
| Solar energy farms | • Perform pre-feasibility studies<br>• Prepare the project development module with relevant business model for setting up of CETP<br>• Implement the possible technology |
| Disaster/hazard management systems | • Prepare the disaster management plans (on-site, off-site) for the identified industries and industrial parks<br>• Design and specify the disaster management infrastructure<br>• Implement the possible technology |
| Waste management | • Includes measures for clearing up wild dump sites and for management of hazardous wastes as well as solid wastes |
| Environment management cells (EMCs) | • Set environment management cells |
| Renovation of infrastructures | • Renovate gray infrastructures to green infrastructure (GI)<br>• Install the natural water circulation system by applying LID/GI techniques |

## 5.4 CONCLUSIONS

The key elements of SDGs include social development, economic growth, and environmental protection aimed to solve economic inequality on a global level, environmental degradation, and the severely damaged ecosystem. Green industry has been adapted as a worthy solution to achieve environmental protection and economic growth at the same time. Green industry is a type of industry based on an eco-friendly system that runs resource and energy efficiency, producing low carbon and waste emission, non-polluting and safe industry through integrated lifecycle management. The EIP is an industrial complex composed of the mechanisms of the ecosystem. EIPs are perceived as important alternatives to build sustainable industrial parks. Industrial ecology for EIPs, which resembles the mass circulation and energy flow of nature, is focused on potential roles of businesses in reducing environmental pollutant loads with managements of the lifecycle of product manufacture. EA and EMA are useful tools that can help the decision-making process in the private sector and actual application of environmental management during the internal decision-making process. MFCA is one of the EMA methods, which is a tool to increase resource efficiency in developing and emerging economies. Ecological engineering is a technological approach for solving various environmental problems in the industrial complex. Recently, a variety of fields such as landscaping, civil engineering, environmental, design, and energy technology have been grafted with ecological engineering in a wide range of industrial complexes. Stormwater management using green infrastructures is one of several practices that can incorporate the ecological engineering concept applicable in industrial areas.

## REFERENCES


Coffman L. (2000). Low-Impact Development Design Strategies, an Integrated Design Approach, EPA 841-B-00-003. Department of Environmental Resources, Programs and Planning Division, Prince George's County, Maryland, USA.

Eco-industrial Park (EIP) (2005). Building Sustainable Industrial Park. Available at: www.eip.or.kr/eng/industrial.html

European Environment Agency (EEA) (2016). Green Economy. Available at: http://www.eea.europa.eu/

Gibbs D. and Deutz D. (2005). Implementing industrial ecology? Planning for eco-industrial parks in the USA. *Geoforum*, **36**(4), 452–464.

Jasch C. (2001). Environmental Management Accounting – Principles and Procedures. UNDSD, New York, USA.

Kokubu K., Kos Silveira Campos M., Furukawa Y. and Tachikawa H. (2009). Material Flow Cost Accounting with ISO 14051. ISO Management Systems.

Leuenberger H. (2015). Implementation of Eco-Industrial Park Initiatives for sustainable Industrial Zones in Vietnam, World Resource Forum.

Lindmark M. (2002). An EKC-pattern in historical perspective: carbon dioxide emissions, technology, fuel prices and growth in Sweden 1870–1997. *Ecological Economics*, **42**, 333–347.

McGranahan G., Jacobi P., Songsore J., Surjadi C. and Kjellén M. (2001). The Citizens at Risk: From Urban Sanitation to Sustainable Cities. Earthscan, London, UK.

Odum H. T. and Odum B. (2003). Concepts and methods of ecological engineering. *Ecological Engineering*, **20**, 339–361.

OECD (2010). Eco-Innovation in Industry: Enabling Green Growth. Organization for Economic Co-operation and Development, Paris, France.

Panayotou T. (1993). Empirical Tests and Policy Analysis of Environmental Degradation at Different Stages of Economic Development, Working Paper WP238, Technology and Employment Programme. International Labor Office, Geneva.

Raskin P., Chadwick M., Jackson T. and Leach G. (1996). The Sustainable Transition: Beyond Conventional Development. Stockholm Environment Institute, Stockholm.

UNEP/CSCP (2007). Raising Resource Productivity in Global Value Chains – spotlights on international perspectives and best practice. UNeP/Wuppertal Institute Collaborating Centre on sustainable Consumption and Production (CSCP), Germany.

UNIDO (2011). UNIDO Green Industry: Policies for supporting Green Industry. United Nations Industrial Development Organization.

United Nations (UN). (2016). Sustainable Knowledge Platform (Sustainable Development Goals). Available at: https://sustainabledevelopment.un.org/sdgs

# Chapter 6

# The restructuring of industrial estates in the Netherlands: The use of a new decision support model for a process analysis of the inner harbor area of Enschede

*Dr. Kjell-Erik Bugge*

*Associate professor, Centre for Urban and Environmental Development, Saxion University of Applied Sciences, k.e.bugge@saxion.nl*

## 6.1 A MAJOR CHALLENGE FOR SOCIETY

The views of how industrial estates should be developed are continuously changing. The 'traditional' approach to the development of such areas in the Netherlands in the past has focused almost exclusively on ensuring that basic infrastructure and facilities for firms were in place. That approach is no longer considered appropriate. Today views suggest that the interests of all stakeholders should be taken into account. Planning agencies take this into account when they try to design and manage interactive processes that can create an 'ideal' situation. The ideal situation, where all stakeholders are accounted for, can be viewed as an industrial estate where an optimum has been found for the 'Triple P': People, Planet and Prosperity (Elkington, 1997; United Nations, 2002).

The search for such ideal outcomes is influenced by several factors. First, the socio-economic and spatial characteristics of specific areas serve as constraints, while at the same time providing specific opportunities. In the Netherlands, for example, areas close to the main national Amsterdam airport, and to the Rotterdam harbor, attract specific firms that see business advantages in the close proximity to these important hubs. At the same time the population density is high in these areas, which, among other factors, means that risks related to external safety will limit the opportunities for some developments.

© IWA Publishing 2017. Wealth Creation without Pollution: Designing for Industry, Ecobusiness Parks and Industrial Estates
Brian D'Arcy, Lee-Hyung Kim, Marla Maniquiz-Redillas
doi: 10.2166/9781780408330_75

A second factor is that the land market in the Netherlands for development of new industrial estates is dominated by local government (Louw *et al.* 2004; PBL, 2009). This situation has led to competition between local governments to attract firms to their industrial estates. The competition has contributed to oversupply of green-fields.

This leads to a third factor, provincial and national governments attempt to develop a 'well-functioning market', which in particular means creating a situation where supply and demand is in balance. For example, the local governments within a region have to co-ordinate and program all industrial estate developments. Green-field development is legally permitted when no suitable solutions for firms in need or more land can be found on the existing industrial estates.

Furthermore, the manner in which the economic situation in Europe has developed over the past 8–10 years has significantly affected views on what can and needs to be done, and what the different stakeholders find acceptable. Firms face serious challenges, and many struggle hard to survive. Plans for growth, and migration to new industrial estates have been cancelled, or postponed. Firms focus on their core business, and any investments in their surroundings (the industrial estate) have low priority or no priority at all. Consequently, in most parts of the country there has been a steep decline in demand for land on new industrial estates. Firms remain situated on the existing industrial estates or, in the worst case, go bankrupt, leaving vacant land and buildings.

As planning agencies, local governments in the Netherlands have a certain responsibility for contributing to maintaining an appropriate quality of existing industrial estates (Alexander & Faludi, 1996; Needham, 2000). However, to save costs and develop a sharper strategic focus on main responsibilities, government organizations at all levels have been downsized and re-organized, forcing fewer people to perform more tasks. This process puts severe constraints on how local governments address city development, and in particular industrial estate development. Although principles and options for achieving a more sustainable development help in the search for solutions, they do not dictate how decisions should be made in specific local situations. There are always partly conflicting interests; always limited resources; always changes that necessitate new searches for additional information and commitment; and always the danger of high ambitions for sustainability gradually 'disappearing' during the development process.

It can be concluded that society faces a major challenge regarding restructuring of industrial estates. Too much, principally public, money has been spent on processes that last too long, results are too often unsatisfactory to current stakeholders and disregard the protection of the interests of the future stakeholders. Here we present an approach for appropriately addressing this challenge. The approach focuses on the use of a new decision support model for achieving quicker interactive processes, and for achieving more sustainable outcomes.

## 6.2 A 'REAL LIFE' EXAMPLE: RESTRUCTURING OF AN INDUSTRIAL ESTATE IN THE HARBOR OF THE CITY OF ENSCHEDE

### 6.2.1 A brief introduction to the city and the industrial estate

The city of Enschede has almost 160,000 inhabitants. It is situated in the eastern part of the Netherlands within the region of Twente. The city used to be a center for the Dutch textile industry until approximately 40 years ago. Textile production was then, within a short period, re-located and taken over predominantly by firms in Asian countries. Since then the employment profile of the city has changed considerably. Today the service industry dominates.

About 22% of all employment in the city is situated on industrial estates (I&O Research, 2014). The case area, the 'Harbor area' industrial estate, is located at the end of the Twente-channel: a channel that connects the city to the main river systems in the Netherlands. The industrial estate is relatively close to the city center. It encompasses an estimated 200 hectares, and accommodates 365 firms employing almost 5600 employees (I&O Research, 2014). The harbor (and channel) is used by various firms for bulk transport; however, the majority of the activities in this area do not use the harbor facilities. The 'Harbor area' is the only industrial estate in the city where firms in the highest environmental risk category are permitted. Risk in this situation is 'translated' into zoning regulations: minimum distance in metres to 'vulnerable objects' such as housing.

### 6.2.2 A growing sense of a need to act, and the first restructuring activities

The 'Harbor Area' industrial estate has been gradually changing over the past decade. Some firms migrated to new estates. The firms that remained offered relatively limited employment opportunities. For some, spatial quality was unimportant for their businesses. Others lacked the necessary means for physical maintenance of their buildings. The effects were a slow deterioration of spatial quality in the area, a decrease in property value, and an increase in extensive land use. This gradual process of decay precipitated an acknowledgement of a need for action, which catalyzed the development of a plan for restructuring the whole harbor area (Gemeente Enschede, 2005). Between 2006 and 2008, several projects were implemented by the local government to manage the restructuring and, in particular, to invest in public property by improvements to infrastructure.

### 6.2.3 Lack of progress and desired results

In order to facilitate integral improvements to the area, involvement and investments of both the local government and the owners of private property are needed.

Co-operation was therefore viewed as a key success factor for the restructuring. In 2009 the co-operation started on the 'Inner harbor': an area of 27 hectares where all land plots are on the waterfront. Original intentions were to alter this section and designate it for use by heavy industry (in this case, firms belonging to the highest category regarding environmental zoning requirements) to make use of the harbor facilities for transport by water. This would create a situation where certain firms would need to relocate thereby improving the harbor facilities, especially the quay. The restructuring was intended to offer new opportunities for accommodation to specific firms. The local government worked closely with the interest group of entrepreneurs in the harbor, assuming that entrepreneurs would only invest in the restructuring if positive effects on continuity and profit were expected for their firms. Unfortunately, the collaborative process did not lead to sufficiently desirable outcomes such as investments in private property. In 2011 the interactive process and progress of the local government and interest group was deemed 'frozen'. In order to facilitate change the local government required more insight into, and an overview of, the process itself. In particular, they needed to identify specific opportunities for changes which could positively affect the situation.

## 6.3 DEVELOPING AN OVERVIEW AND AN UNDERSTANDING: A QUESTION-BASED PROCESS DECISION SUPPORT MODEL

### 6.3.1 Complex decision-making taking place in arenas

Decision-making processes for the restructuring of industrial estates in a Dutch context are complex, with many actors involved. Each actor has different views of the problems defined and different opinions on the preferred solutions. The decision-making in such situations usually takes place in 'rounds' (Teisman, 1992) within 'arenas' (Ostrom, 2005). An 'arena' is a combination of involved actors and a specific challenge. Each round in the arena proceeds through the steps of searching for solutions, assessing their desirability and feasibility, and making the actual choices needed to make things happen. The goal of each round is to create a situation where a sufficient number of actors are committed to the implementation of the selected solution (Ostrom, 2005).

### 6.3.2 A process model for supporting practitioners in their decision-making

There are many options for analyzing and influencing the interactive processes within an arena. Process models are particularly suitable for handling a large number of variables at the same time (Vrolijk, 1996). The analysis of the process in the harbor in Enschede was performed using a new process decision support model (Bugge, 2013, 2015), developed for the purpose of improving industrial estate restructuring. It views the interactive processes in a restructuring as taking place within two consecutive arenas. First, there is the initiative arena. In the Netherlands, a planning agency, typically the local government, works towards getting sufficient commitment from all key stakeholders to start a joint initiative for improving the industrial estate. To reach this goal sufficient *information* about the situation on the industrial estate, the identification and involvement of *motivated* preferred participants, and sufficient *resources* for taking the next step, are all required. When sufficient commitment from all the stakeholders has been reached, the renewing process enters the second arena. During this arena, the implementation arena, plans are moved towards the selection and implementation of specific improvement measures. Again, the same three key variables are addressed: the *motivation* of the actors to invest, the *information* about problems and improvement options, and the availability of *resources* for implementing the selected solutions. Or, in short: the actors must know what to do, and be willing, and able to do so (Figure 6.1).

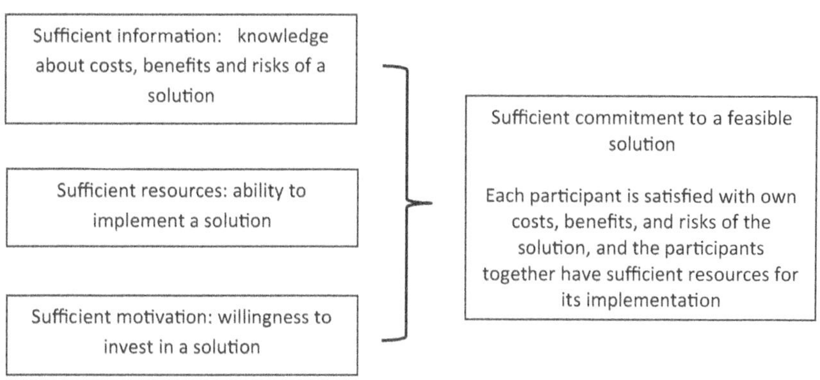


**Figure 6.1** Commitment relies on sufficient motivation, information, and resources.

The model was developed to meet the needs of practitioners. Consequently it is designed to be accessible, transparent, and compact (Bugge *et al.* 2010). To meet this objective the model uses four main questions for the implementation-arena which are consecutively addressed, and each question must be answered with a positive answer in order to proceed:

(1) Does the planning agency know enough about the vision(s) of the stakeholders?
If not, how can it gather the necessary information (i.e., for identifying the initial conditions)?

(2) Does the planning agency know who the preferred participants are?
If not, how can it gather the necessary information (i.e., for identifying the initial conditions)?

(3) Are the preferred participants willing to spend time and money (process costs) on working out the vision into an action plan?
If not, how can they be influenced (i.e., for identifying the initial conditions)?

When one or more action plans have been worked out:

(4) Is there sufficient commitment from the preferred participants to spend (capital) resources on implementing one of the action plans?
If not, how can that be influenced (i.e., managing the action arena)?

Each step works according to an 'if … then' approach. *If* a question leads to a negative answer, *then* follow-up actions are suggested. Such actions can be specific in-depth follow-up questions, analyses of the interests and goals of stakeholders, analyses of risks or feasibility of possible improvement measures, or an organizational re-design. Actions are always aimed at identifying the needs and the level of satisfaction of the participants related to a specific development. They are used for influencing motivation for investing in the process and its outcomes, and for finding solutions having acceptable distributions of costs, benefits and risks among the participants.

## 6.4 THE APPLICATION OF THE DECISION SUPPORT MODEL TO THE INNER HARBOR AREA OF ENSCHEDE

The goal of the process analysis was to improve the understanding of the complexity of the restructuring of the industrial estate. The improved understanding was to provide specific opportunities for bettering the restructuring and facilitate a continuation of the process.

### 6.4.1 Initial design should focus on interests and involvement of actors

Obstacles: the first challenge was to 'translate' the process model into a workable design for this specific process analysis. A 'detached' approach, only analyzing documents, was not suitable, since it was important to develop a shared understanding of how to proceed, which requires an interactive approach. That works as a process intervention: it influences the way the participants think and act. Taking into account the project constraints, a 'lean' design was chosen consisting of three interactive sessions. Each session focused on collecting information and working toward conclusions about how to proceed. Working on conclusions so early in the process reflected an increase in shared views and commitment. The design needed to be adaptable, allowing for changes in agenda and participant needs.

Consequently, design choices for these sessions were initially limited to three. First, it was decided to work with small groups (6–7 participants), using a combination of 'brainstorm' and discussion. This enabled open interaction, and ample opportunity for contribution to the process from each participant, making a small group more efficient than a larger group.

Secondly, choices were made about the composition of the groups for each session. The participants of the first session were limited to representatives of the local government. Representatives of both entrepreneurs and the local government were invited to the next two sessions. This approach was chosen because the local government can be seen as a 'complex actor': it consists of several departments that focus primarily on different areas of policy. It was important to start the interaction with the entrepreneurs based on a complete overview of information and opinions of the different internal actors. Finally, the main topics for the agenda and the phases (see Figure 6.2) were chosen.

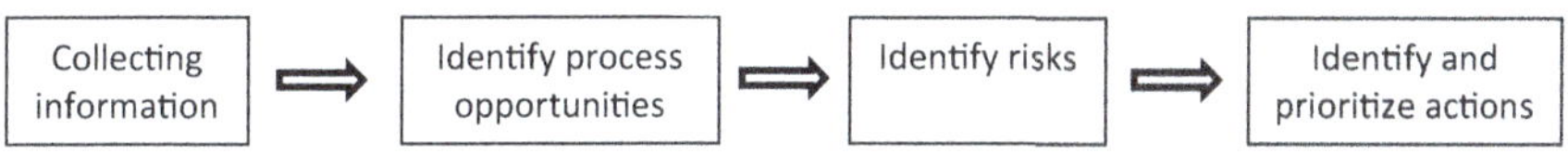


**Figure 6.2** The main phases and activities in the process analysis.

Taking into account an anticipated need for intermediary adaptations, the objectives were only specified for the first session. This session focused on identifying the availability, quality, and distribution of information within the organization of the local government. In particular, the session was intended to provide information about the current situation on the industrial estate: interests, objectives, availability of resources, and plans and the expected willingness of stakeholders (both departments of the local government and entrepreneurs) to invest in the process and specific measures to make it feasible. It was expected to lead to preliminary ideas about process opportunities, process risks, and uncertainties.

### 6.4.2 Outcomes of the first session: Rich information, but lack of completeness and overview

The outcomes of the first session were quite 'rich'. The participants provided a large variety of detailed pieces of information about the situation on the industrial estate. In particular, they described the high ambitions for the site development (usually referring to the responsibility of the local government), and how such ambitions have been translated into a large number of plans and actions. These activities focused on factors such as adequate infrastructure, efficient use of space, sustainable development (including energy efficiency and reduction of carbon dioxide emissions), employment, and quality of life issues.

On the other hand, the participants described how the planning process has faced several major set-backs, and how this may be explained by a mismatch in planning processes. These dynamics were nicely illustrated by one of the participants, saying: 'When the local government was ready to take action, then the commitment of the entrepreneurs was diminished, and when the entrepreneurs wanted to take action, then the local government had to start the decision-making'.

The (text) analysis of the detailed information further revealed six problems regarding access to information:

(1) Insufficient clarity about any shared vision of the local government on the development of the Inner Harbor area.
(2) No structured and complete overview of information (data) for the individual firms in the area regarding satisfaction with location factors and plans for the future is available.
(3) Inadequate transparency concerning political ambitions related to harbor developments.
(4) Not clear whether the firms in the area experience the existing problems as their own problems: i.e., accept problem 'ownership'.
(5) Unknown if commitment is available for joint actions by local governments and firms.
(6) No complete overview of situation regarding compliance with laws and regulations.

The first five problems all refer to topics related to commitment. A provisional conclusion was therefore that the lack of information represented potentially serious process risks, including for identifying process opportunities.

### 6.4.3 Bridging the sessions: Redesign according to need

It was expected that the representatives of the entrepreneurs could fill in some of the information gaps. A second objective was to explore and identify their commitment to ideas for follow-up developed during the first session, although they had to be modified to accommodate altered plans which emphasized the development of harbor-related activities (development of more quays, as well as attempts to influence the migration of firms into, and out of, the area, and providing opportunities for firms making use of, or having plans for, transport by water).

### 6.4.4 Outcomes of the second session: Commitment remains the key issue

The change of plans regarding development of the harbor would have significant physical consequences for the infrastructure and design of the area. It was common knowledge that entrepreneurs profiting from the developments would support the new plans. On the other hand, it was unknown to what degree active opposition would be encountered from the firms not profiting from the developments. The key challenge was therefore to develop a strong business case. New solutions needed to be found, and for each solution acceptable distributions of costs, benefits and risks among the involved stakeholders was needed. Thus, the second session appeared to indicate that 'everything' must be done all over again. This was however, by no means the case. The results underpinned the fact that the participants of the session acknowledged the importance of appropriately addressing the information problems identified during the first session.

### 6.4.5 Bridging the sessions: Addressing changed priorities

The third session was designed to address issues arising from the changed plans, and the possible divergence of opinions between the first and second sessions. The participants were first individually asked to identify how to address process opportunities (the outcomes are presented below). A three-step approach was used. The risks (probability × negative effects) related to not addressing process opportunities were identified. Subsequently, follow-up actions were suggested and prioritized. Finally, a short discussion, focused on the outcomes and their consequences for a process toward reaching a shared view on how to proceed, was conducted.

### 6.4.6 Final outcomes

The outcomes of the final session strongly reflected the effect of the changes to the plans for the area. The main conclusion was that it was clear that insufficient commitment for implementation of the plans was available. This was understandable, taking into account that the plans had been changed quite recently. On the other hand, the stakeholders preferably involved in the process were known. For some of these stakeholders their willingness to invest was also known. The session led to a prioritization of follow-up actions. The high priority actions were:

- *Create a 'shared view' on the restructuring*: All stakeholders needed to know, and acknowledge, aims and constraints for the restructuring, and they needed to be actively involved in scenario-development.
- *Develop a 'map' of process opportunities and process risks*: The complexity of the restructuring is high. The possible effects of specific measures and solutions are accordingly difficult to assess. Developing an easily adaptable and dynamic 'map' of the area would enable all stakeholders to get an overview of pros of cons of measures.
- *Develop new, and make use of emerging, 'golden opportunities'*: A 'golden opportunity' is a combination of a good idea (solution), and sufficient commitment and resources for its implementation. The process management of the restructuring should aim at designing such opportunities, while at the same time also spotting and using temporary 'opportunity windows' (e.g., caused by co-financing). This implies influencing and monitoring all three factors. In particular, restructuring program-managers can be given a clear mandate for making decisions within a pre-defined 'range' regarding costs and effects on policy fields.

- *Ensure appropriate combinations of specific challenges and specific skills*: Interaction between representatives of firms and local government in a restructuring covers a large variety of topics and skills. In particular, the restructuring program management should ensure that involved personnel possess the necessary skills for development of commitment and business cases such as motivation and negotiation.
- *Perform an integral compliance-check on laws and regulation*: Restructuring is a way to address the process of decay. One of the reasons for decay is that firms pay limited attention to how their activities affect their surroundings. In some cases this may result in non-compliance with spatial and environmental laws. An integral compliance-check can reveal such problems. Appropriate actions can be taken to ensure that quality is maintained in the restructured area.

More specific actions, which would contribute to the above mentioned high-priority actions, were also recommended. These actions were aimed at improving access to information about the firms, and influencing progress and quality of outcomes of the restructuring. The first recommended action focused on how firms tend to target on their 'own' problems. Firms often perceive problems as limited to their own areas (plots of land) and factors that directly affect continuity of the firm. Problems that affect the performance of the whole industrial estate are either 'neglected' or perceived as something that should be solved by, for example, the local government. It was therefore recommended that how individual firms perceive the necessity for joint action to improve the industrial estate be identified, as well as to what degree they would be willing to contribute to joint actions. In particular, it was vital to establish whether individual firms would stimulate or 'block' specific developments. All these topics are to some extent 'sensitive', because the information has strategic value. It was therefore recommended that a representative of the firms be used for collecting the required information. By doing so, there could be no discussion of the intentions of a representative from the local government. The preferred representative needed therefore be well-known and trusted by the other entrepreneurs in the area, and at the same time, be neutral towards the outcomes.

Information on a variety of other topics was also needed. Trends, objectives and plans, agreements, and measures taken and their effects on site performance needed to be evaluated. It was therefore also recommended that a central data-base be developed that would include all information relevant to the restructuring. This data-base should, in particular, contain fact-sheets on all individual firms. A regular updating of the information is necessary, and it was therefore recommended that a platform be created for information exchange involving representatives of firms and the local government.

## 6.5 LESSONS LEARNED

Which lessons can be learned from the application of the decision support tool in Enschede? A first observation is the process analysis involved interacting with people. It was therefore important to understand why the participants acted in a group setting the way they did. The participants involved in this process analysis represented their own organizations. At the same time, they were expected to take into account the effects of the restructuring on the future of the whole industrial estate and even the local community. The participants also had a common 'history'. They had met earlier on several occasions related to the restructuring, but also related to other city-development projects. The specific 'history' of the restructuring had led to the starting point and perceived necessity of the process analysis: an almost 'frozen' co-operation process between entrepreneurs and the local government. Earlier opinions were all known and all participants most likely had biased views of each other regarding issues such as skills, motives, and trustworthiness.

### 6.5.1 1st lesson learned: Focus on motives, and do not forget the impact of a 'common history'

An important challenge was therefore to try to 'separate facts from feelings', and to create an open interaction based on mutual trust. The first session was therefore limited to participants representing different policy fields within the organization of the local government. It was assumed that 'starting small' would stimulate an open exchange of information about the perceived reasons for the problems. This worked well. The information output was rich, and a common view on how to proceed in the next sessions was established. However, using this approach was not without risks, because in the second session the entrepreneurs were 'confronted' with a view on how to proceed. They could be suspicious as to why they were excluded from the initial process, with a risk of reduced trust and a related lack of willingness to participate. The lesson learned was that this dilemma can best be addressed by remaining transparent in discussion, while maintaining a clear focus on progress and goals. All output of the first session was therefore introduced in the second session as draft conclusions and recommendations. Agendas and participants of the sessions were also kept abreast of changes. This approach also created opportunity for all participants to re-evaluate the desirability of the intermediary outcomes with key stakeholders of the organizations they represented. Further, working in this manner made it easier for the participants to express ideas and opinions without fear of saying something that could cause permanent damage.

### 6.5.2 2nd lesson learned: Balance the need for progress and the need for open discussion

Despite the positive effects of this open process there were also clear time-constraints to be taken into account. In particular, this applied in the interactive sessions. The experienced time-pressure ensured that participants recognized the need for working efficiently. Time was well spent. Even the time spent on discussing the strategic change of plans for the industrial estate did not cause any serious set-back to progress.

### 6.5.3 3rd lesson learned: Use time-pressure as an incentive for efficiency and to-the-point outcomes

The design of roles for performing the process analysis was another important factor. In this case the researcher was responsible for both moderating the sessions and analyzing their outcomes. This was an efficient solution, but brought with it the risk that the results could be influenced. In practice, the approach turned out to be the best option for two reasons:

(1) Lack of trust is an important issue in 'frozen' processes. An intermediary was needed who could create and maintain room for an open dialogue, while at the same time making participants focus on the future instead of the past. Furthermore, 'objectivity' was required: a 'third party' that could be trusted to perform the analysis without being influenced by any personal bias related to the outcomes.

(2) The limited time available and the complexity of the topics. The moderator had to continuously monitor and guide the interaction to keep focus on key issues. The moderator needed to ensure that all the main questions necessary for the analysis were answered, and at the same time try to identify interesting patterns of relationships between factors and variables that could be explored by the group.

### 6.5.4 4th lesson learned: Combining the roles of researcher and moderator: effective, but not without risks

Participants were not always focused on the full complexity of the issues at hand. Practitioners are used to work according to project management principles. In decision-making they tend to apply 'reductionist' approaches, focusing on one issue at the time, trying to translate all complex decisions into a number of specific ('smaller') decisions. Each decision preferably addresses a business case that is easy to assess. The decision-support model needed to be simple to use and at the same time allow 'translation' of the rich detailed information into aggregated answers to the main questions on, for example, commitment to investments.

### 6.5.5 5th lesson learned: Adapt decision-support to fit the specific situation

*In summary: Success in restructuring depends on appropriate process design and process management*

A design needs to specifically combine the right agenda, participants, and the right goal-oriented facilitated interaction at the right moment. Restructuring is about investments in measures that improve the area. The combination of complexity (many topics to be addressed at the same time) and long-lasting processes makes it difficult to develop shared commitment to such investments. The strategic change to restructuring objectives that occurred during the process analysis, illustrated the importance of the effects of change: the agenda, the group of participants, and the way of working in the sessions had to be adapted. Creating shared commitment will always be the main challenge for a successful restructuring. This is clear in the Dutch restructuring context.

## 6.6 SOME BRIEF REFLECTIONS ON FUTURE NEEDS FOR A MORE SUSTAINABLE DEVELOPMENT OF INDUSTRIAL ESTATES

Commitment reflects priorities; regarding restructuring, priorities have changed significantly in the Netherlands, since the economic events of 2008. 'People, Planet and Prosperity' remain the key issues, but the (re)development of industrial estates has been confronted with new challenges. Creating more employment opportunities is a main objective on a national level. Firms however, focus on continuity and have fewer resources available for measures that do not have a direct effect their business.

Local governments cannot continue investing in restructuring on such a large scale as was the case in the last decade. They have to reprioritize and reduce their efforts. Firms have to accept a more substantial part of the shared responsibility for the quality of the industrial site. An emerging trend is therefore the search for new public-private co-operation approaches that fulfill these requirements.

One of the characteristics of the Dutch approach to industrial estate development has been the lack of an effective link between 'brownfield' and 'greenfield' development. Too many new estates have been developed, stimulating migration of firms from existing sites to new estates which has often led to undesired negative effects such as vacant plots.

A second issue has been the development of 'master plans'. In the past, such plans often served as rigid frameworks for future developments. One provisional conclusion is that a co-operation model needs to be a 'growth-model', designed for flexibility and to continuously take into account the needs of all stakeholders. Such a model should include much more than the 'traditional' management issues such as maintenance of the area and collective services. It should focus on the 'Triple-P' topics (People, Planet and Prosperity), and it should involve firms and local government in a joint continuous improvement approach. Recent developments suggest a shift within the 'Triple-P balance'. Entrepreneurs are increasingly trying to find well-working approaches to the combination of 'people' and 'prosperity', for example creating employment opportunities (using specific co-funding). Firms and the local community seem to increasingly understand their need for each other. In that respect economy-driven necessity stimulates a more holistic approach to (re)development of industrial sites.

Finally, it is clear is that no investments will be made unless all the benefits of the business case are satisfactory for all those involved. Decision-support will remain an important instrument for developing commitment, and interaction is the way to build a 'bridge' between individual motivation and shared commitment. Within the Dutch planning tradition, there is a need for alternative, better interaction models that appropriately address the altered needs and opportunities of both firms and local governments.

## REFERENCES

Alexander E. H. and Faludi A. (1996). Planning doctrine: its uses and implications. *Planning Doctrine, Special Issue of Planning Theory*, **16**, 11–61.

Bugge K. E. (2013). Restructuring Industrial Sites more Quickly: A Decision Support Model. PhD thesis, Radboud Universiteit Nijmegen.

Bugge K. E. (2015). Besluitvorming herstructurering bedrijventerreinen. In: De markt voor bedrijventerreinen: uitkomsten van onderzoek en beleid, E. van der Krabben & C.-J. Pen (red.)

Bugge K. E., Welter F., O'Gorman W. and Hill I. (2010). Regional sustainability, innovation and welfare through an adaptive process model. In: Facilitating Sustainable Innovation through Collaboration: A Multi-Stakeholder Perspective, J. Sarkis, J. J. Cordeiro and D. Vazquez Brust (eds), Springer, pp. 77–96.

Elkington J. (1997). Cannibals with Forks: the Triple Bottom Line of 21st Century Business. Capstone, Oxford.

Gemeente Enschede (2005). Masterplan Havengebied Enschede. Gemeente Enschede, Enschede.

I&O Research (2014). Monitor Economische Ontwikkeling Enschede 2013. I&O Research, Enschede.

Louw E., Needham B., Olden H. and Pen C.-J. (2004). Planning van bedrijventerreinen. Den Haag, Sdu Uitgevers.

Ostrom E. (2005). Understanding Institutional Diversity. Princeton University Press, Princeton.

PBL (2009). De toekomst van bedrijventerreinen: van uitbreiding naar herstructurering. Planbureau voor de Leefomgeving, Den Haag/Bilthoven.

Teisman G. R. (1992). Complexe besluitvorming: een pluricentrisch perspectief op besluitvorming over ruimtelijke investeringen. VUGA, Den Haag.

United Nations (2002). Report of the World Summit on Sustainable Development, Johannesburg, South Africa, 26 August–4 September 2002. United Nations, New York.

Vrolijk H. C. J. (1996). Methoden en Regels voor de Projectplanning: Een Toepassing van Beslissingsondersteunende Systemen. PhD-thesis, University of Groningen, Labyrinht Publication, Capelle a. d. IJssel.

# Chapter 7

# Eco-innovation opportunities in the waste management sector in Scotland

*Juliette O'Keeffe*

*Division of Environmental Engineering and Systems Visualisation, University of Abertay Dundee, Bell Street, Dundee, DD1 1HG, E-mail: j.o'keeffe@abertay.ac.uk*

## 7.1 INTRODUCTION

In the transition to low carbon and more sustainable economic models, one of the primary challenges facing future waste and resource management systems is the shift away from a flow of materials from use to final disposal, to a system that retains and maximises the use of resources within the economy. The circular economy approach recognises that the supply of global resources is finite and dwindling, and therefore seeks to maximise the use of resources in a circular 'beginning to beginning' approach as opposed to a linear 'beginning to end' approach by optimising the lifespan and multiple uses of a single resource over time. The World Economic Forum and the Ellen MacArthur Foundation have produced a comprehensive review (World Economic Forum, 2014) of the economic potential of a circular economy (CE) and estimate that over US$1 trillion in cost savings could be made by 2025 by adopting new approaches to resources throughout the supply chain. However, such approaches would require participation by a wide number

© IWA Publishing 2017. Wealth Creation without Pollution: Designing for Industry, Ecobusiness Parks and Industrial Estates
Brian D'Arcy, Lee-Hyung Kim, Marla Maniquiz-Redillas
doi: 10.2166/9781780408330_89

of groups across the economy including the waste and resource management sector, manufacturing, product design, consumers and energy producers. Although these models envisage a global coming together of participants across the economy, practical and feasible interactions are also possible on a local scale. The ***eco-industrial park*** concept is one model for applying industrial symbiosis on a local level. The concept, and its application in Scotland, is discussed further in the following sections.

### 7.1.1 About industrial symbiosis

As described by Boons *et al.* (2011), industrial symbiosis (IS) refers to '*the material and energy flows and transformations that are generated by economic actors within a geographically bound system*'. Generally, IS involves the exchange of by-products of processes between co-located entities for mutual benefit. Chertow (2008) states that there are three areas where resource exchange may take place between industry: by-product exchange; utility or infrastructure sharing; and joint provision of services. Chertow distinguishes IS from normal exchanges, for example between a single organisation and a recycler by suggesting that in order to meet the definition of IS, three or more organisations should be involved in the exchange of two or more resources. This can include energy, heat or materials with the result being the net minimisation of waste and resource consumption. The IS approach can assist organisations in meeting regulatory requirements in areas such as waste disposal, carbon reduction or water effluent release.

Generally the characteristic of co-location and geographical proximity is required for IS to occur, as noted by Boons *et al.* (2011). The extent of this geographical boundary however is not well defined. Jensen *et al.* (2011) suggest that for simple materials exchanges, the definition of 'in proximity' may depend on '*the spatial distribution of industrial diversity*' in the region. Jensen *et al.* (2011) mapped out exchange distances for 792 material exchanges between organisations participating in the UK's National Industrial Symbiosis Programme (NISP) to December 2009. They found that across all exchanges there was an average resource movement of 20.4 miles, suggesting that the extent of industrial diversity in the UK is approximately 20 miles, the distance across which at least half of all industrial exchanges should be possible (Jensen *et al.* 2011). Geographical proximity of a much-reduced distance however is necessary if the symbiosis involves the sharing of infrastructure and services. Such exchanges measured by NISP accounted for approximately 187 of 979 identified exchanges. These included synergies such as sharing of land, labour, or logistics, many of which must occur within a much closer proximity. These types of synergies in combination with material exchanges can form the basis for an eco-industrial park development.

### 7.1.2 The eco-industrial park concept

Industrial parks provide ideal locations for IS opportunities to be developed, particularly where industry involved in manufacturing, recycling, or energy production is located in close proximity. These locations provide opportunities not only for material exchanges, but also for shared services and infrastructure including surface water treatment, process and wastewater treatment as well as sharing of expertise, labour and logistics. The application of IS on an industrial park level, where many participants are involved in multiple exchanges of resources, may be termed an eco-industrial park (EIP). Although various definitions exist to describe an EIP, it is generally understood that there is a geographical grouping of businesses that, as defined by Lowe (2001) *'seek enhanced environmental, economic and social performance through collaboration in managing environmental and resource issues'*. Therefore multiple IS exchanges between businesses on the park are likely to be occurring and additional features may exist such as sustainable buildings, pollution control measures, shared services and sustainable infrastructure (Lowe, 2001; Chertow, 2008). Lowitt and Cote (2013) suggest that the definition of an EIP can range from a vision based on efficient use of materials and energy to one that takes a broader view of incorporating additional factors such as the local ecology, water and air management. Lowe (1997) describes a number of strategies for developing an eco-industrial park including: incorporating features such as onsite waste management functions; designs that maximise the efficient use of energy and water, and incorporate renewables and eco-friendly materials; systems for communications between organisations and managing joint aspects of the park such as maintenance of infrastructure and provision of services.

The development of EIPs can be a spontaneous and organic process built up by tenant businesses on an industrial estate, or can be pre-planned from a defined vision. The most commonly cited example of IS applied on an industrial-park scale is that of Kalundborg, in Denmark, where a coal-fired plant exchange of heat, steam, fly ash and waste gypsum with a range of other co-located organisations was not pre-planned. The EIP emerged over time as the need to find outlets for industrial by-products for a large coal-fired plant and the availability of cheap resources encouraged company interactions (Lowe, 1997; Chertow, 2008). This was a gradual development based on business need, however planners, authorities and businesses are now exploring ways of how to either build new EIPs from a pre-envisioned network of symbiotic connections or retrofit existing industrial parks with new IS interactions in order to deliver economic and environmental benefits (Behera *et al.* 2012). The remainder of this paper reviews the outcomes of a European Regional Development Fund (ERDF) project delivered in Tayside, Scotland to explore the possibility of building EIPs in the region. The case study illustrates some of the difficulties developers will experience both in attempting to retrofit an EIP concept onto an existing industrial park, and creating a new park.

## 7.2 CASE STUDY: INDUSTRIAL SYMBIOSIS AND ECO-PARK DEVELOPMENT IN EAST SCOTLAND

### 7.2.1 Background to ACE eco-partnerships

At a European Union (EU) level, environmental legislation has been driving change in all member states to improve environmental quality and reduce carbon emissions. Directives such as the Water Framework Directive, the EU Landfill Directive, and the Renewables Directive (2009/28/EC) have led member states to introduce new regulation in order to contribute to achieving EU wide targets. Scotland has declared ambitious targets for improving water quality (Water Environment and Water Services (Scotland Act 2003)) reducing emissions (Climate Change Scotland (Act) 2009), and diverting waste from landfill (Waste (Scotland) Regulations, 2012). Therefore the policy environment for businesses seeking to improve their environmental impacts has been supportive, including government funding being made available for support agencies and development projects designed to stimulate opportunities for industrial symbioses. The ACE Eco-Partnerships project, hosted by the University of Abertay Dundee, received part funding through the ERDF in 2008 to investigate opportunities for industrial symbiosis and the potential to create an EIP in the Tayside and Fife region of Scotland. The project investigated both the retrofit of an EIP on an existing park and the development of a new EIP from a pre-envisioned collection of businesses.

### 7.2.2 Scoping for retrofit EIP development potential on existing industrial estates

The difficulty in applying the EIP concept to existing industrial estates, is the lack of initial planning, and siting of tenants and lack of pre-defined industrial symbioses that exist on a new EIP development.

The ACE Eco-Partnerships project carried out scoping studies of industrial estates in the east of Scotland (Tayside and Fife) to identify those that could benefit from the application of an EIP concept, and where potentially synergistic relationships between tenants already existed. In order to do this, the project consulted local maps of industrial estates showing existing tenants, consulted economic development information for the locality, and examined potential environmental issues that could be overcome more effectively by building partnerships between the tenant businesses. Site visits were also carried out to identify important features of the site for consideration in evaluating EIP development potential. Categorisation of business types by sector or activity was carried out to identify the types of waste production likely across the site, the level and type of energy demand (space heating, process), any manufacturing present and the types of resource inputs, and other resource or environmental management issues that may be present.

Using this information, the project team nominated industrial estates that would be most likely to benefit from a collective approach to environmental issues such as waste, energy or water treatment, for further investigation and study. Businesses were contacted and offered a free resource efficiency audit in order to entice them to participate. Those businesses that accepted were visited and interviewed about their existing levels of resource consumption and wastage and any additional environmental matters such as process emissions to air or water. This process identified one of the primary difficulties with retrofitting the EIP concept onto existing industrial sites. Although there was some uptake by businesses interested in resource efficiency audits, the majority of businesses were either unwilling or uninterested in participating. Those that did participate were struggling either with waste and energy costs or environmental compliance of some nature. Although these businesses were interested in identifying improvements to their own site and processes, there was much less willingness to consider joint measures with other tenant businesses. Another issue encountered by the project team was the high percentage of site tenants that did not own the building or the land where they were operating. This reduced interest in investing in updates or improvements to infrastructure that would have potential longer-term environmental benefits. The owner-occupier businesses also showed some reluctance to enter into joint agreements with non-owner-occupiers, whose tenancy on the industrial estate was seen as potentially transient. An issue faced by factor organisations or site owners was that any improvements to infrastructure to reduce existing environmental impacts would be based upon the current tenancies, however these were not fixed.

Despite these difficulties, there was some willingness amongst participants to consider where working with others could have a mutual benefit, particularly in areas of waste management – where economies of scale could be provided for a specific waste stream, or the opportunity to recycle a specific waste stream could only be realised if working in partnership with others. In addition, the correct mix of site owners and business types can have a marked effect on whether the businesses may consider working in a more symbiotic relationship. A high concentration of long-term tenants, coupled with businesses involved in manufacturing or processing are better suited to EIP retrofit as compared to sites where large numbers of business units show regular turnover in tenants and where businesses are primarily linked to retail, wholesale or distribution activities. Where the local authority provided support for co-ordinating or highlighting opportunities to local businesses, there appeared to be greater support from businesses to consider EIP development opportunities.

### 7.2.3 Scoping for potential EIP on new sites, or single tenant sites

The ACE Eco-Partnerships project carried out a review of a small number of identified sites, where an existing anchor tenant was interested in exploring opportunities to develop an eco-industrial park, to propose how an eco-park development could progress. Three locations were identified as potential EIP locations where either limited or no economic activity was currently taking place. Each location was visited by the project team and discussions focused on the potential redevelopment of the sites into EIPs utilising the theory of industrial symbiosis.

A site evaluation was carried out for each site, which considered factors of relevance for eco-industrial park development such as the geographical location of the site and existing linkages to other businesses, main transport routes, and connection to electricity and gas grid and sewer connections. Characteristics of the site also helped to identify the relevant features and technologies that could be incorporated on the site, such as potential renewable energy technologies, or requirements for contaminated land or effluent treatment. Proximity to existing large industry was also considered (potential partners in developing a site), as well as proximity to residential neighbours potentially impacted by a development (i.e., due to noise, odour, vehicular movement, or as an end-market for resources such as heat).

#### *7.2.3.1 Site A*

Site A was a former industrial site that had been vacant for approximately five years. The site owner was interested in the EIP concept, and planned to redevelop the site to include a range of economic activity. The 42-acre site was previously used for industrial activities over the past two centuries but at the time of the site investigation was somewhat isolated in terms of proximity to other active industrial activities. The site benefited from close proximity to transport linkages including major road networks connecting to Scotland's central belt and the potential for a rail connection nearby, in addition to proximity to the River Forth. The site also had connection to mains services and an existing water treatment facility onsite. The site owner was interested in the types of IS synergies that could be developed, with the potential to create added value partnerships by the careful selection of tenants. In particular, the site owner was involved in waste collection and recycling, and was therefore interested in the potential industries that could co-locate to manufacture products from recovered waste materials. In addition, the potential to generate energy onsite from waste using anaerobic digestion of organic waste materials was of interest.

A number of promising strategies were discussed with the company, including the potential for businesses involved in reprocessing or remanufacture, as well as other potential tenants with high heat demand that would benefit from access to heat from an anaerobic digestion plant. Despite a number of potential strategies being proposed and initial enthusiasm by the developer, a number of barriers to EIP development existed. The proximity of neighbouring residential properties could potentially restrict some of the industrial activities that could take place due to noise, odour or other nuisance issues. However, as the site had been an industrial facility for the past two centuries at least, it was likely that this barrier could be overcome. In addition, the proximity of residential properties could potentially be an asset as an end user of renewable heat or electricity generated onsite.

In order for redevelopment to occur, the developer would have to consider the site contamination in great detail, with the possible need to remediate some areas of ground, and possibly asbestos from existing buildings. Many of the structures onsite would needed to be demolished or refurbished in order to be suitable for future tenants. Unfortunately the site suffered a large fire in December 2012 that has affected further development. The costs of remediating and rebuilding on the site have delayed any further progress at this point.

#### *7.2.3.2 Site B*

Site B was a former military site with limited existing structures, however with ample space for development, close to Perth. The site owner was interested in the EIP concept however, with no existing activities taking place on the site, any proposed EIP would be based on attracting a suitable collection of tenant businesses. As the site was a blank canvas, the potential for development ideas was endless. However, without an existing anchor tenant, or identified demand from industry for occupying the space, defining the types of businesses and potential synergies that could take place on the site was difficult. In order to consider how to attract businesses, a review was undertaken to recommend features of a development that would encourage businesses to relocate. This included features such as shared resource and waste management facilities and collection, potential interconnection of units to allow sharing of resources, and an overall eco-friendly design to the development incorporating green infrastructure features to minimise the impact of future development and business activities on the environment.

The difficulty in proposing an EIP design was in identifying the types of businesses likely to co-locate and the features that would allow other businesses to interact in a synergistic manner. Any development on the site would potentially suffer from a lack of existing tenants. In order for industrial synergies to occur, the site owner would be reliant upon potentially synergistic businesses seeking tenancy in the industrial park. Although encouragement could be provided in terms of infrastructure, or discounts/incentives to the 'right' kind of businesses, there would be no guarantee that any business would choose to locate in the site. This approach to eco-industrial park development attempts to pre-determine the types of IS that may emerge and seek out potential businesses that match the profile. This can impose limitations on the development, and assumes all potential actors will participate in developing collective solutions.

#### *7.2.3.3 Site C*

Site C was an existing waste management site, located near to agricultural land, with an existing landfill and composting facility onsite. The site owners employed a consultant/project manager specifically to explore the idea of developing an EIP on the site. The presence of an existing waste management facility and composting facility, including food waste recycling, onsite helped the consultant to identify a number of potential synergistic businesses to target for co-location and the beginning of an EIP.

The location of the site between Fife and Perth was relatively remote, with little potential for impact upon residential areas. The site itself had experienced land contamination issues in the past related to the operation of the waste management facility, and was still regularly monitored for impacts on the local aquatic environment and vegetation. There was some evidence that the site could benefit from improved infrastructure, including features to protect the aquatic environment from pollution impacts related to business activity on the site.

The most promising area to begin developing EIP synergies on the site was in relation to the processing of food waste. It was observed that the high value compost from the composting of food waste could be utilised locally as a medium for local food growing, particularly high value food items to be grown in green houses. The site also has significant potential for energy generation from a range of sources that the site owner was considering including wind, anaerobic digestion and a waste to energy thermal plant. The production of energy onsite would be particularly beneficial if tenants could utilise the energy directly rather than export offsite. This was particularly true for heat, which led to exploring the potential for co-locating industry with a heat demand (such as greenhouse food growing). In addition, as a waste management facility, there was also potential to segregate materials such as plastics in order to provide a feedstock to complementary manufacturing businesses to establish production units.

The site is currently still in the process of forming ideas and attempting to attract complementary businesses to co-locate. ACE Eco-Partnerships assisted the developer in exploring the requirements and complexities of many aspects of the development, however what appears to be limiting the development is the willingness for additional businesses to co-locate. Although the site may provide a convenient location for onsite energy generation, food production and potentially manufacturing from recycled waste, the location and lack of existing infrastructure for business units limits the attractiveness of the site to potential tenants.

### *7.2.3.4 Summary*

Table 7.1 presents some of the advantages and disadvantages of each site with respect to possible EIP development.

**Table 7.1** Advantages and disadvantages of EIP development by site.

| Site | Key Advantages | Key Disadvantages |
|---|---|---|
| A. Ex-industrial site | – Large anchor tenant involved in materials reprocessing<br>– Good connection to services<br>– Large space and convenient location for additional businesses to co-locate | – Cost of remediation<br>– Cost of redevelopment of buildings<br>– Proximity to residential properties<br>– Attracting interested companies to co-locate |
| B. Former military site | – Large space and convenient location<br>– Ability to incorporate green infrastructure from the outset | – No existing tenants, therefore no pre-defined synergies to develop<br>– Making an EIP 'fit' would depend on willing businesses co-locating and agreeing to work synergistically |
| C. Waste management site | – Large remote space<br>– Supply of waste materials to supply recycling or manufacturing activities<br>– Ability for renewable energy development onsite to provide potential tenants with low carbon heat and electricity | – Remoteness of site may limit attractiveness to some businesses<br>– Inability to attract complementary businesses<br>– Existing pollution and contamination issues, and need for improved infrastructure |

Although EIP development potential existed on all three sites visited, establishing a new EIP is a complex process. Where an anchor tenant already occupies, or owns the site, it is easier to begin to envisage potential synergies and the types of businesses that would benefit from relocation to the site. There are limitations with these sites however including location and attractiveness of the site to new tenants. The promise of access to resources in the form of feedstock or energy may be attractive, but businesses may also wish to be in locations that are accessible and appealing to potential customers. For a new site with no existing tenants, a major limitation in establishing EIPs is finding synergistic businesses that are looking to relocate at the same time. It is difficult to pre-envisage the types of synergies likely to occur, and simultaneously identify businesses willing to relocate and take up tenancy in newly designed eco-parks.

## 7.3 DISCUSSION

The ACE Eco-Partnerships project example has highlighted some of the difficulties in establishing new EIPs. There are a number of theories as to why any individual eco-industrial park project may not succeed. Boons and Spekkink (2012) define five factors that are relevant to creating ideal conditions for industrial symbiosis to occur:

(1) Learning and strategic vision
(2) Issues related to the diversity of actors involved
(3) Trust
(4) Anchor tenants or co-ordinating bodies
(5) An enabling context (i.e., policy, regulation, structures, culture etc.)

Boons and Spekkink (2012) analysed these factors in the context of institutional capacity, including relational capacity, knowledge capacity and mobilisation capacity and found that relational capacity (strengthening the network collaborators and partners, building trust and evaluating risks and opportunities) and knowledge capacity (awareness of the opportunities for IS and what is realistically feasible, and what tools and resources may be available) were not significantly linked to successful IS developments. Mobilisation capacity, however, (the ability of actors to engage with externals, influence policy or regulation, and to draw in resources from external sources) was significantly linked to successful industrial symbiosis measures. This theory is supported by the observations made in the ACE Eco-Partnerships project. Of the three sites, only Site C could be observed to have strong mobilisation capacity. Through the ACE Eco-Partnerships project, the relational capacity and knowledge capacity of organisations was improved, however, without the ability to act on these aspects of improved institutional capacity the developments are yet to go ahead.

Interestingly both Site A and Site C were initiated by waste management companies. In the context of current waste policy and thinking on the circular economy, a waste management firm as an anchor tenant for an EIP could be a valuable institution. However, waste firms focused more on waste collection and transport rather than sorting, processing and developing partnerships with end-markets may not be fully subscribed to maximising the benefits of IS. While the supply of recyclate is increasing, the capacity to process materials within Scotland has not fully developed (O'Keeffe & Gilmour, 2012). Closing resource loops in Scotland remains challenging with large quantities of recyclate being exported out of Scotland and often out of the UK. Eco-industrial park developers seeking to improve mobilisation capacity will need to address this issue and lobby government to stimulate end-markets for waste. In order to achieve this, issues related to quantity of recyclate, the availability of processing facilities and technologies and end markets for recycled content products need to be addressed (O'Keeffe & Gilmour, 2012).

## 7.4 CONCLUSION

Eco-industrial park development has the potential to improve the efficiency of industrial operations from waste management to manufacturing to energy production, and provide added benefits to businesses in terms of reduced resource management costs and improved environmental credentials. However, developing EIPs requires willing participants, the opportunity for synergistic interactions, and the ability for actors to implement the necessary interventions to allow synergies to develop. Although EIP development is largely business led, local authorities can assist the process by providing local businesses with information and examples of best practices and, on local authority owned sites, support collective solutions that support business and protect the environment.

## REFERENCES


Behera S. K., Kim J., Lee S., Suh S. and Park H. (2012). Evolution of 'designed' industrial symbiosis networks in the Ulsan Eco-industrial Park: 'research and development into business' as the enabling framework. *Journal of Cleaner Production*, **29–30**, 103–112.

Boons F., Spekkink W. and Mouzakitis Y. (2011). The dynamics of industrial symbiosis: a proposal for a conceptual framework based upon a comprehensive literature review. *Journal of Cleaner Production*, **19**, 905–911.

Boons, F. and Spekkink W. (2012). Levels of institutional capacity and actor expectations about industrial symbiosis. *Journal of Industrial Ecology*, **16**(1), 61–69.

Chertow M. R. (2008). "Uncovering" industrial symbiosis. *Journal of Industrial Ecology*, **11**(1), 11–30.

Jensen P., Basson L., Hellawell E., Bailey M. and Leach M. (2011). Quantifying 'geographic proximity': Experiences from the United Kingdom's National Industrial Symbiosis Programme. *Resources, Conservation and Recycling*, **55**, 703–712.

Lowe E. A. (1997). Creating by-product resource exchanges: strategies for eco-industrial parks. *Journal of Cleaner Production*, **5**(1–2), 57–65.

Lowe E. A. (2001). Eco-Industrial Park Handbook for Asian Developing Countries. Indigo Development, Oakland, CA.

Lowitt P. and Côté R. (2013). Putting the eco into eco parks. *Journal of Industrial Ecology*, **17**(3), 343–344.

World Economic Forum (2014). Towards a Circular Economy [online]. Available from: http://www3.weforum.org/docs/WEF_ENV_TowardsCircularEconomy_Report_2014.pdf

# Chapter 8

# Green industrial park practice: A case study of green infrastructure in Wenling, China

*Nian She[1], Jiang Zhao[2], Minhua Zhao[3] and Wei Wu[2]*

*[1]Shenzhen University*

*[2]Zhenjiang Urban Planning and Design Institute*

*[3]Shanghai Urban Hydrology and Hydraulic Planning Research Institute*

## 8.1 INTRODUCTION

During the past 40 years China's urbanization has grown rapidly. The urban population increased from 17% in the late 1970s to 56% in 2015, faster than any country in world history. However, the rapid urbanization has huge environmental impact to the country: more frequent urban flooding, severe air pollution, heat island effect, and deterioration of rivers and lakes. In 2012, a heavy storm hit the capital city, Beijing, that caused the death of more than 70 people and a billion dollars of losses in property damage. Therefore the State Council in 2013 issued No. 23 and No. 36 documents that required every city to upgrade its drainage network in the next decade to prevent urban flooding from 20- to 100-year storm events.

In October of 2014 the federal government embarked on 'sponge city' initiatives. Shortly after that 16 cities were selected for a 'sponge city' pilot construction program. Each city was awarded $63 million dollars per year for three years. In April of 2016 another 14 cities were selected as the second pilot group. According to China's State Council the total investment in China's sponge city construction will be trillions of dollars in the next 15 years. In October of 2015 the State Council

© IWA Publishing 2017. Wealth Creation without Pollution: Designing for Industry, Ecobusiness Parks and Industrial Estates
Brian D'Arcy, Lee-Hyung Kim, Marla Maniquiz-Redillas
doi: 10.2166/9781780408330_99

issued an official document (No. 75) that requires more than 20% of the urban area of all cities to meet the objectives of the 'sponge city' initiative by the year 2020 and 80% by the year 2030.

A sponge city is a city that acts as a sponge with an urban environment planned and constructed to soak up almost every raindrop and capture that water for reuse (Geiger, 2015; Yu, 2016). The concept is quite similar to the United States' Low Impact Development (LID), the United Kingdom's Sustainable Urban Drainage Systems (SUDS) and Australia's Water Sensitive Urban Design (WSUD) (Stahre, 2006; D'Arcy, 2013). In this paper we will describe how a small city in Zhejiang providence named Wenling, integrated LID with the conventional conveyance system in its new economic development zone, before the initiative of 'sponge city'.

When the new economic zone was in the planning phase, the question of how to build the urban drainage system was raised by the city government and Zhenjiang Urban Planning and Design Institute in 2013. The debate was focused on whether or not the city should adopt LID as part of its drainage system. The planning and design institute and some government officers were advocating for a green stormwater infrastructure (GSI) solution because the water shortage in the new economic zone is severe, but many engineers and administrators were in favor of a huge gray infrastructure by building big stormwater trunk sewers, huge detention tanks and powerful pumping stations. Both sides have their merits: since GSI was new in China then, no one had construction experience, and there were no construction manuals, guidelines and regulations. More importantly, there was little data to support the performance of GSI in China at the time. Therefore, it was very risky to implement LID citywide. Another concern was that LID/GSI cannot handle large storm events, especially high intensity-short duration storm events. To overcome these barriers the planning and design institute had studied many cases in the United States and two successful examples in China – Shenzhen University LID demonstration project and Longmu Bay LID application in a sub-tropical island in China. After a long debate the city decided to adopt LID as an important component of its urban drainage system. The design and construction started in 2014 just prior to the 'sponge city' initiative.

## 8.2 BACKGROUND

Wenling is a small city of 926 square kilometres located at the coast area of the Eastern China Sea, with more than 1.2 million population. Its gross domestic product (GDP) per capita is about $7756, ranked 21st among the top 100 small cities nationwide. To improve the city's competitiveness, a new district named Wenling Eastern New Development Zone was built as an industrial park to attract investment from high tech and manufacturing companies worldwide. The administrative officers and planners envision that the new district should not only be the industrial and tourist center, but also an ecologically sound livable town. However, the challenge to the ecological development is the lack of water.

Although the precipitation in Wenling is abundant, the city cannot take advantage of it because it is located in a valley surrounded by hills, and no large water reservoir can be constructed nearby. The river flow passes through the city rapidly during the wet season, sometimes causing severe erosion. But after the storm the river banks are dried out due to steep slopes. This means that during the wet season flooding is a risk, but during the dry season water supply is at risk.

## 8.2.1 Water shortage – the bottleneck for the development of Wenling new district

The new district is about 37 $km^2$ (Figure 8.1). It is land-filled from the sea. The precipitation of the area is 1600 mm, but 95% of runoff is 'let go'. It is not economically feasible to build a large reservoir for water supply, since the regional geological condition is complex, and the land resource is limited. Based on the forecast of the master plan, the demand for freshwater will be 770,000 t/d by the year 2020. The majority of the water supply will come from desalination. The average freshwater in Wenling is about 696 $m^3$ per capita which is far below the national average of 2200 $m^3$ per capita (Figure 8.2).

**Figure 8.1** Arial Photo of Wenling's new district before the landfill. The white boundary is the land-filled new district.

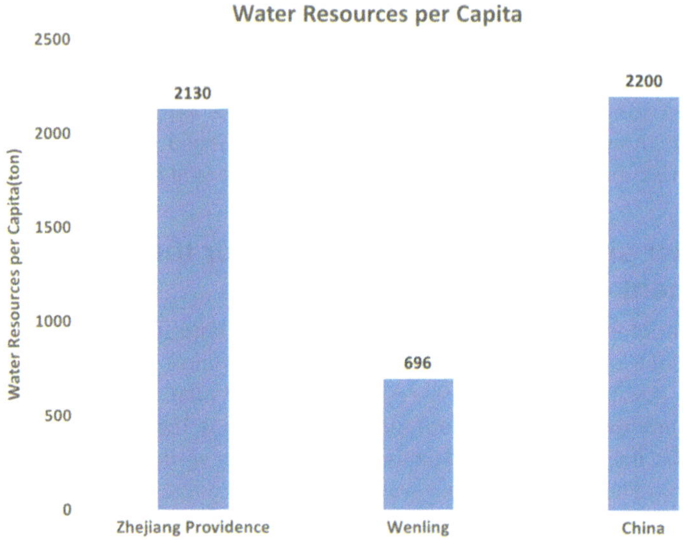


**Figure 8.2** Water resources in Wenling comparing to the province and nation.

## 8.3 A SOLUTION – GREEN STORMWATER INFRASTRUCTURE

To conserve the freshwater resources while accomplishing economic growth, the planners, engineers and developers worked together to develop an ecologically sound growth master plan for the next decade. It requires the city and businesses to design and install GSI that allows rainwater to infiltrate into the ground close to where it falls, and make use of soil and plants to filter and absorb stormwater. Sound GSI design involves the preservation of natural areas, especially those in low-lying places that naturally collect stormwater, and the grading of the landscape to disperse runoff from roofs, roads, and parking areas into existing natural areas or specially designed bioretention facilities (Stahre, 2006; D'Arcy, 2013; Liu & She, 2013; Geiger, 2015; Yu, 2016). In areas where pavement is required, strategies are used to minimize the area of pavement, and, in some cases, permeable paving materials are used that allow stormwater to infiltrate (Liu *et al.* 2015; She *et al.* 2015).

To encourage businesses and developers to conserve the freshwater, the district government passed a regulation that allows the water utility to use water fees as a leverage tool for water management. The regular water fees for the businesses, industries and developers are 10 RMB/ton ($1.6 USD/ton), but if the business and industries utilize the rainwater, they can obtain some tax deduction and cover their investment in less than 10 years. The water conservation programs include wastewater reclamation, rainwater harvesting, and low impact development (LID). For the new development it is required to use LID to mitigate stormwater runoff to mimic natural hydrology, such that the hydrologic characteristics of post-development are almost the same as those of pre-development. In particular, it is required to use a combination of bioswales, tree boxes and detention ponds to convey and treat runoff up to a 10-year storm event for all roads and sidewalks (Chen *et al.* 2015; Li *et al.* 2015; Wu *et al.* 2015).

## 8.3.1 Detention facilities

For manufacturers in the industrial park, no stormwater from their campus is allowed to discharge directly to the drainage system. It must go through a detention facility for treatment (Figure 8.3). In this way, the inspectors can visualize the water quality of the detention facility so that the illegal discharge of industrial wastewater through storm drains can be observed and corrected easily. Since this enforcement, no illegal discharges through storm drains have been found in the park.

**Figure 8.3** Detention ponds in the industrial park. All stormwater from the campus must be discharged to the ponds.

## 8.3.2 Bioretention

Since the traffic on the major roads is heavy, if using conventional drainage pipes to convey the runoff alone the pollutants generated from the traffic would wash off to the receiving water near the shorelines. This would cause damage to the fragile habitats. Therefore, a bioswale is required for every new constructed road. The runoff must pass through the bioswale for infiltration and purification before discharge to a conventional drain at the end of the road. In general, no drainage pipes are installed along the road. The benefits of installation of bioswales are significant. They not only save the cost of drainage construction and prevent pollutants from the heavy traffic entering the receiving water, but also reduce the volume and peak flow of runoff to prevent flooding of the road. In July of 2015 a typhoon hit the new district. A total of 100 mm of rain fell on the land within a short time period. The government officers were worried about road flooding interrupting heavy traffic.

It was a surprise that no water was ponded in the bioswale. Figure 8.4 shows the bioswale operation just after the rain storm stopped.

**Figure 8.4** Bioswale in Jintang Road right after a 100-mm storm event.

Bioretention is also applied to the campus as rain gardens. Figure 8.5 shows the rain gardens in Union Zhilun Manufacture. The downspouts of the roofs are disconnected, and the runoff from the roads are also routed into the rain gardens for treatment.

**Figure 8.5** Disconnection of roof downspout and rain garden.

Figure 8.6 shows the layout of the rain gardens at Jinhong Food Processing Equipment Company.

The company campus is about 3.67 ha, 60% of the campus area is the buildings roofs. All roof runoff is either treated by the rain gardens or by the bioswales. The treated runoff is then routed to the detention pond for recreational uses. During a heavy storm the pond can be used as additional storage.

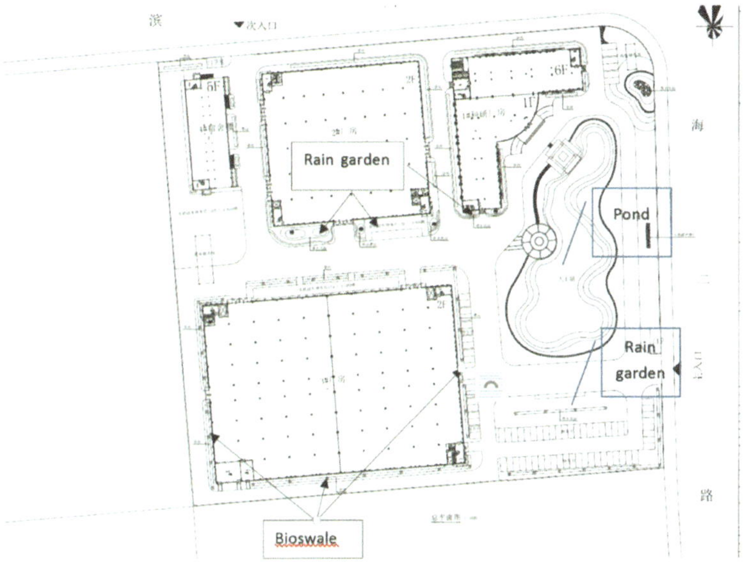


**Figure 8.6** Bioretention layout of Jinhong Food Processing Equipment Company.

Figure 8.7 below shows one of the rain gardens in the district. The design and construction of the rain garden were completed by the company itself. It should be noted that an infiltration trench was added to the rain garden. The treated runoff is all conveyed to the pond.

**Figure 8.7** Rain garden and infiltration trench in the campus.

### 8.3.3 Rainwater harvesting

Since the incentives for water conservation are attractive, many companies in the district voluntarily invest in rainwater harvesting. For example, Jinhong Food Processing Equipment Company harvests rainwater for cooling buildings. Others harvest rainwater for restroom flushing, and irrigation. Figure 8.8 below shows a rainfall harvesting facility. The left figure shows a pre-treatment settlement tank, where the debris and suspended solids are removed. The right figure shows the underground storage. Pre-treated runoff passes through a filtration system before going into storage. The harvesting system is designed and built to recycle the rainwater in the summer time. The water is pumped from the storage tank to the roof top. A sprinkle system spreads the water to form a fine water curtain that absorbs heat from the roof top. The spread water is collected by roof drains and flows back to the pre-treatment settlement tank. The harvesting system also has an overflow outlet. The water level is controlled by a weir. The harvesting system is shown in Figure 8.9.

**Figure 8.8** Rainwater harvesting system: pre-treatment and storage tank.

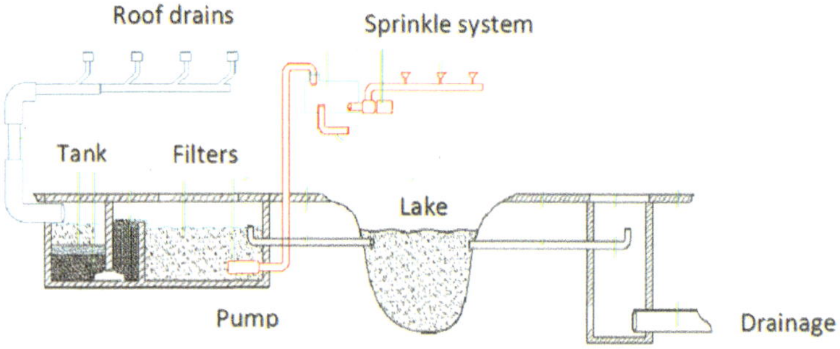


**Figure 8.9** Rainwater harvesting system.

## 8.4 CONCLUSIONS

This case study shows that green stormwater infrastructure can be integrated with conventional conveyance systems to solve water shortage and urban flooding problems. The significance of this case study is that if the government sets up an attractive incentive policy the businesses and developers are willing to conserve the water voluntarily. The benefits of GSI are obvious, in that it saves money by replacing the drainage network along roadways; prevents urban flooding during a heavy storm event; eliminates illegal industrial discharge to urban drainage systems; and saves energy cost during the summer and increases the aesthetic values of the properties.

From Wenling's GSI practice we have solved an important problem that has concerned many cities in China for years – illegal industrial wastewater discharge to the municipal drainage system. By enforcing stormwater discharging to an open detention pond any illicit connection or illegal discharge will be easily observed by the inspectors. Hence correction or law enforcement can be undertaken quickly.

It is encouraging to us that many businesses and developers are willing to accept GSI as their campus's infrastructure construction. Some of them even designed the GSI themselves under the guidance of a local design institute. Some GSI innovation products are also seen in this case study. We believe that the Wenling case study can provide significant experience to other cities.

## REFERENCES

Chen B., Liu J., She N. and Xu K. (2015). Optimization of Low Impact Development Facilities in Beijing CITIC Complex. International Low Impact Development Conference 2015, Houston, Texas, January 19–21.

D'Arcy B. J. (2013). Managing Stormwater: from aspirations to routine business. *gwf-Wasser*, International issue 2013, 32–38. www.wassertermine.de

Geiger W. F. (2015). Sponge City and LID technology – vision and tradition. *LAF (Landscape Architecture Frontiers), 014*, **3**(2), 10–21.

Li T., Shan S., Liu J., She N., Chen B. and Wu L. (2015). Applying New Features of Low Impact Development Techniques in the Master Planning of Guangzhou Educational Town. International Low Impact Development Conference 2015, Houston, Texas, January 19–21.

Liu J., Lucas W., Chan I., She N. and Wu L. (2015). Innovative Design of the LID Facilities of Congdu Ecological Grange. International Low Impact Development Conference 2015, Houston, Texas, January 19–21.

She N., Liu J., Lucas W., Li T. and Wu L. (2015). The Performance of LID Treatment Train in Shenzhen University during Extreme Storm Events. International Low Impact Development Conference 2015, Houston, Texas, January 19–21.

Stahre P. (2006). Sustainability in Urban Storm Drainage – Planning and Examples. Svenskt-Vatten, Stockholm. ISBN 91-85159-20-4.

Wu L., Lucas W., Liu J., She N., Xiao D. and Ye G. (2015). Innovative Design of the Gravel System for Underground Parking Garage. International Low Impact Development Conference 2015, Houston, Texas, January 19–21.

Yu K. (2016). Sponge City: Theory and Practice. Proceedings of the International Conference on Green Infrastructure and Resilience, 16–19 March 2016. The Water Environment Federation (WEF) and Ecological Urban Planning Academic Committee of Chinese Urban Planning Society, Shenzhen University, Shenzhen.

刘建、佘年：第一届低影响开发技术国际交流及技术培训会议演讲资料汇编，1-181页，第一届低影响开发技术国际交流及技术培训会议，中国深圳，2013年12月4-6日.

# Chapter 9

# Drainage infrastructure for industrial and commercial premises, estates and business parks

B. J. D'Arcy[1*], N. McLean[2], L.-H. Kim[3], and C. Pittner[2]

[1]Independent environmental consultant

[2]WSP chris.pittner@wspgroup.com

[3]Kongju National University, Korea

*Corresponding author: brian@enviroexperience.co.uk

## 9.1 INTRODUCTION

Following the Industrial Revolution in the UK, industrialised towns and cities developed where resources (e.g., coal, iron ore) and transport for raw materials and exports (e.g., coastal ports such as Glasgow, Liverpool and London) favoured the establishment of processing and manufacturing industries. Severe pollution problems resulted (Porter, 1973). Drainage was, until the 1960s, typically a combined sewer, so that all potential sources of pollution from industry, as well as greatly varying stormwater flows, played havoc with treatment plant operations whilst often almost un-noticed at the source factories.

From the 1960s in the UK, policies to establish employment opportunities more widely in towns and cities resulted in the creation of industrial estates or trading estates, where industries such as light engineering, printing, assembly, distribution, and many others were attracted to locate and develop their businesses. Relatively few had large trade effluent discharges, and those that did often discharged to foul sewer and treatment. The drainage infrastructure provided by then was usually a separate sewer system, but at most premises the potentially polluting activities took place on the hard-standing areas at the back of factory buildings, where only surface

© IWA Publishing 2017. Wealth Creation without Pollution: Designing for Industry, Ecobusiness Parks and Industrial Estates
Brian D'Arcy, Lee-Hyung Kim, Marla Maniquiz-Redillas
doi: 10.2166/9781780408330_109

water drainage was provided, creating pathways for pollution. Once again, pollution resulted (D'Arcy & Bayes, 1995; Scottish Environment Protection Agency [SEPA], 1999). Equally seriously for industry, it became clear that the designation of an area of land as an 'industrial estate' did not mean that the infrastructure had been designed with consideration of the needs of industry moving onto the site. Parallel developments, and environmental problems, in many countries led to the publication of a report by the United Nations Environment Programme (UNEP, 1997). The guidance in that report was very limited with regard to stormwater management, although progressive and useful in other respects (e.g., eco-industry ideas of sharing wastes to utilise them as raw materials, clustering businesses to facilitate that, and providing shared services where possible). The UNEP (1997) report does specify permeable pavement for staff car parks, and use of wetlands or retention ponds, but indicates in the guidance on-line features in a watercourse, and aside from the car parks, does not advocate source control for stormwater management. Heal *et al.* (2005) and Ibanez (2015) have highlighted the risks of pollution in end-of-system industrial ponds without a treatment train of upstream measures. The UNEP (1997) report also has guidance on emergency planning, but does not indicate benefits of such considerations in designing the drainage infrastructure.

So, in designing infrastructure requirements for industry how can water pollution risks be managed and minimised for environment and for business?

The strategic approach taken here to planning the drainage infrastructure for industrial and commercial properties involves two essential aspects:

(a) Incident risk management and contingency planning
(b) Designing to capture and treat diffuse pollutants in runoff.

Happily, (a) and (b) are compatible, and when both are considered, the resulting drainage infrastructure should be fit for purpose, as indicated below. A third category of pollution risk is the misconnection of trade or sewage effluents into the surface water system; that is partly a procedural issue, but use of open SUDS techniques (see below) can make such errors less likely or at least more obvious.

## 9.2 SUSTAINABLE DRAINAGE TECHNOLOGY

### 9.2.1 Bringing various objectives into a single technology

The use of passive landscaping features for pollutant capture and treatment (variously known internationally as structural Best Management Practices (BMPs), Low Impact Development (LID) or sustainable urban drainage systems, SUDS), allows for management of pollution risks, which must include contingency planning when applied on industrial/commercial premises. In addition, of course, the infrastructure must meet basic requirements to drain the premises, and also address flood risks. The idea of designing features for managing quality and quantity aspects of stormwater runoff, but also addressing a multiplicity of additional opportunities in the creation of drainage features, has been described as sustainable drainage (D'Arcy, 1998, 2013). The variety of positive parallel benefits if a feature is optimised in that way include well-being for local people (the workforce at a factory for example), wildlife habitat creation, informal recreation (walking around or alongside features), bird-watching, picnic opportunities away from the office for lunch breaks, physical exercise in getting out and walking, etc. All those positive attributes of creating a stormwater feature have been encapsulated in a loose use of the term 'amenity' in Figure 9.1 below. The term *sustainable urban drainage systems* (SUDS) developed from that concept, and has found favour with engineers and planners in many countries (e.g., Stahre, 2009; Casal-Campos *et al.* 2012; Hreggviosdottir & Amadottir, 2016).

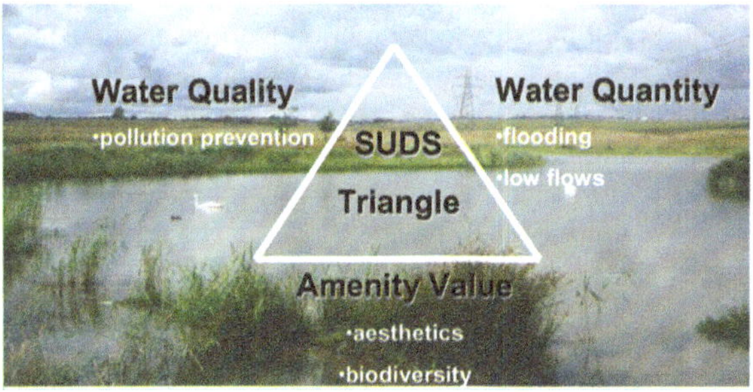


**Figure 9.1** The sustainable drainage triangle concept – designing for multiple benefits (after D'Arcy, 1998).

## 9.2.2 SUDS in relation to industrial estates

This section provides an introductory guide to types of techniques available when an architect, landscape architect or engineer is considering a new development or redevelopment, for industrial/commercial premises or estates. Table 9.1 lists selected measures, and indicates the scope of the various types of features for meeting the drainage and pollution risk management requirements of SUDS infrastructure for industrial/commercial premises. Introductory descriptive details are given in the text below, together with references to detailed guidance. Several general comments can be made:

- Where soils allow infiltration, risks of groundwater contamination are an important consideration. Equally, on contaminated land the impervious layer capping the contaminated material must not be disturbed (a shallow source control swale requires less depth than conventional pipes for example). Where soils favour infiltration but it is a risk and not desired, then an impervious liner is required, or a different technique.
- Where mown grass is required, for example for aesthetics, slopes should not exceed 1:4.

- Filtration through the grass requires far gentler slopes, for example, 2–5% (CIRIA, 2015)
- Open vegetated features allow site management and staff to more easily see evidence of bad practice or accidents and thereby improve site management and understanding of drainage systems and pollution risks (Boogaard *et al.* 2014).
- Where features are provided for industrial/commercial yards, peripheral locations are desirable and bollards or large kerbs (with scope for water movement between them) may be necessary.
- An alternative to flat or nearly flat grass areas for filtration, is to use wet-base features, where uneven ground can enhance biodiversity and save costs grading the surfaces, provided inlet and outlet weirs are designed to achieve flooding of the base for sedimentation at least in part. Examples include sediment forebay and outlet micropool in extended detention features (Schueler *et al.* 1992), and wet swales operating as linear wetlands (Wilson *et al.* 2004).
- Flow controls are necessary to optimise performance of the SUDS features (flow attenuation and pollutant removal).

**Table 9.1** Summary attributes of example types of SUDS features for industrial/commercial sites.

| Feature | Positive Attributes | Drawbacks | Comments |
|---|---|---|---|
| Grass filter strip | Filtration, biodegradation | Land take | Need wide flat space |
| Grass swale | Good connectivity option | Land take | Under-drained swales reduce width |
| Permeable pavement | No land take | Most systems unsuitable for trucks | Reduces nos. pipes & gullies |
| Gravel filter strip | Small land take | Maintenance | Grass cover perpetuates porosity |
| Bioretention units | Small-scale, modular | Limited capacity | Drainage planters |
| Green roofs | Clean-up atmospheric pollutants, attenuate small rainfall events | May be structural requirements to support the roof | Biodiversity benefits, also insulation and extend life of a flat roof |
| Green walls | Scope for innovation and retrofits | Uncertainties for sustained performance | Rarely designed for stormwater management |
| Detention basin | Effective site control | Land take | Good site control |
| Retention ponds | 3–4 weeks residence time | Land take | V. good regional control |
| Stormwater wetlands | Long residence time | Land take & maintenance | Pre-treatment SUDS vital |

## 9.2.3 Example SUDS features

The following details are generally in accordance with UK guidelines (CIRIA, 2015), and additional viewpoints from other sources are each referenced (examples of each are given in Figures 9.2–9.11). Comments and the selected features have been included specifically for industrial/commercial applications.

### 9.2.3.1 Grass filter strips

*Gently sloping grass area on prepared sub-base receiving runoff from immediately adjacent surface*

- Overall gradient typically 2–5% sloping away from hard surface.
- Length of filter strip (from hard surface to down-hill edge) should be greater than 2.5 m; lengths of 5 m have been shown to provide good water quality treatment were vegetation is dense.
- Requires sheet flow across full extent of grassed area.
- Top soil depth of 150 mm on top of engineered sub-grade of 300 mm minimum.
- Filter trench collection may be useful below the grass strip (collection) as flow spreaders at the head of the features, and in larger features on slopes up to 20%, as measures to control flow and achieve successive lower gradients.

**Figure 9.2** *Left*: Grass filter strip taking car park runoff, Germany. *Right*: Filter strip draining to conveyance swale, J4M8 distribution centre, Livingston, UK (*photo: Chris Pittner*).

#### 9.2.3.1.1 Advantages

Effective treatment is achievable. Open space is often required in new developments and filter strips can provide this whilst delivering a passive treatment function. Most maintenance is as for other grass landscaped areas. Visual appearance is similar to other green space areas. Reinstatement may be necessary after exceedance events.

#### 9.2.3.1.2 Disadvantages

Land take and flat land can be difficult, and large areas might be at risk of later redevelopment uses. No significant storage volume for managing flood risks, or storing fire water. Maintaining laminar flow across a grass strip can be difficult in practice, especially if liable to high solids loadings.

### *9.2.3.2 Grass swales as source control measures*

*Shallow linear depression creating an informal channel in grass*

- Shallow grassed channel with gently sloping sides and a flat bed. A gentle fall is achievable if check dams are used to allow successive step-down along a slope.
- A trapezoidal profile favours linear flows, minimising turbulence and hence erosion of bed or sides.
- Gently sloping sides allow use as a source control feature, providing filtration of suspended matter in runoff from a road or industrial/commercial yard for example.
- Sloping flat kerbs at edge of hard surfaces ideal to feed runoff into a swale and allow for some grass turf growth
- If used as part of a pipe-free system, should be at worst cost neutral, by comparison with pipes and gullies.

**Figure 9.3** UK examples: *Left (Photo credit Neil Mclean)* Swale with sheet flow inlet over level kerb from road surface (Forth Valley, Stirling). *Right*: Grass swale with check-dams at Stephen's Croft, Lockerbie (timber industries development, 1990s).

#### 9.2.3.2.1 Advantages

A major advantage on a commercial yard is that casual disposal of oil or other pollutants will stain or kill a patch of grass and be obvious to supervisors and management. That way cleaner practices and recovery of value from waste can be encouraged. Lesser levels of contamination by oil are also effectively treated by soil-grass systems since they favour biodegradation *in situ* (Horner *et al.* 1994; Napier *et al.* 2009; Leroy *et al.* 2016).

#### 9.2.3.2.2 Disadvantages

Land take is an issue on a small site, although under-drained swales will require less space. Care needs to be taken to ensure the feature is not damaged by vehicle wheels straying into it.

Management needs to understand the functions and importance, and not replace swales with pipes or build over them.

### 9.2.3.3 Grass swales as conveyance features

*Vegetated channels, flowing in wet weather, connecting drainage from successive sites*

- Side slopes ideally 1:4 or gentler, but can be 1:3 if no source control filtration function, and stability and practicality for mowing are not issues. Floor of swale generally 0.5–2 m wide.
- Wet-base, natural vegetation often colonises with increasing distance from source areas.
- Under-drained swales can be dry and reduce land take.
- Check-dams needed if the slope over length of swale exceeds 3%; successive check-dams can allow swales to be used on sloping ground as steep as 1:10.

**Figure 9.4** Conveyance swales, Clyde Gateway Industrial Estate, Scotland, UK.

#### 9.2.3.3.1 Advantages

Should be cheaper to install than conventional pipe-and-gully drainage systems, although greater land take. A major advantage is easy access for interventions in the event of a traffic accident in an industrial area.

#### 9.2.3.3.2 Disadvantages

If there are any conventional drains upstream on an estate or business park, then a conveyance swale usually has to be excessively deep to sustain an adequate fall into it. Land take is an issue on a small site. Risk of piped replacement to free up space as redevelopment/growth occurs if within industrial premises.

### 9.2.3.4 Bioretention

*Shallow vegetated landscape depressions which treat runoff from impermeable areas using engineered soils and vegetation*

- These features are also known as biofilters and raingardens (FAWB, 2009).
- Detention depth on surface of up to 150 mm, to allow time for infiltration.
- A prepared soil or sand-based filter medium of good depth – minimum 400 mm, but more typically 700–1000 mm.
- A 'transition' layer, which may be a 100 mm coarse sand, or a geotextile, which prevents washing of sediments and fines and retains filter medium in place.
- Drainage layer at floor of system collects runoff via stone fill into perforated pipe and outlet.
- By providing engineered free-draining subsoils and filter media, the capacity of a green infrastructure feature to accept runoff and reduce pollutants in runoff is enhanced.
- Biofiltration is a popular technique in parts of the USA and Australia, particularly for retrofit, due to modest land take.
- Features may be soft-edged, soil-&-vegetation features wholly integrated into a green landscape, or packed as hard-edged units fitted into highly built-up contexts (at site boundaries, car parks or corners on industrial yards).

**Figure 9.5** *Above left*: Bioretention feature, Melbourne Docklands, Australia. *Right*: Bioretention features taking car park runoff, Portland, USA.

#### 9.2.3.4.1 Advantages

Can be incorporated into normal landscaping and can be especially effective as green infrastructure components in areas that would otherwise be hard landscaping. Bioretention areas or cells can also be very appropriate in car parking and yard areas to provide demarcation of certain areas, for example, edge of property boundary, parking or turning zones, etc. Applying several small cells can be easier to allocate than one large area such as a pond or basin.

#### 9.2.3.4.2 Disadvantages

These systems need the correct permeability of filter media to allow adequate infiltration through layers. There are maintenance fears for the complexity of multi-layer structures such as bioretention features. Blockages will result in greater velocities passing over the surface of the system, short-circuiting it and preventing adequate treatment. The outlet from the system will be at depth which in turn may lead to connection difficulties depending on levels on the site.

### 9.2.3.5 *Green roofs*

*Green roofs (or living roofs) are vegetated across most of, or the entire, roof with several benefits*

Green roofs can capture rainfall and attenuate runoff, the extent of which will depend on several factors including the depth of substrate and drainage reservoir layer and the pitch (steepness) of the roof (CIRIA, 2015; Arias *et al.* 2016). Two main categories:

- Extensive green roof;
  - 20–150 mm depth of growing medium or substrate.
  - Will support simple, low maintenance vegetation, often supporting lesser biodiversity potential.
- Intensive green roof;
  - At least 150 mm depth of growing medium (substrate).
  - Greater depth allows a more extensive habitat and therefore greater biodiversity potential.
  - If there is employee access to the roof then gives amenity value.

**Figure 9.6** *Left*: Extensive green roof, Charles de Gaul airport. *Right*: Restoration work on intensive green roof at MONA (Museum of Old and New Art), Hobart, Tasmania.

#### 9.2.3.5.1 Advantages

Green roofs have the potential to capture rainfall and allow a slow release over time to minimise flood risk, and there are no land-take requirements for a developer. This is especially useful during short, intense rainfall which would otherwise runoff directly to receiving drainage networks and watercourses, increasing flood risk. Additional benefits include improved air quality, extended life of the roof system, reduction of urban heat island effect through evaporative losses and reduction in ambient air temperature. Green roofs can also offer significant habitat and biodiversity potential and can provide thermal and sound insulation. Evapotranspiration cooling effects can reduce air conditioning use and therefore the energy and carbon performances of the building.

#### 9.2.3.5.2 Disadvantages

Due to the increased loading on the roof structure, additional strengthening will usually be required. This may be reduced where modern construction specifications require high thermal mass installations as standard. Access for maintenance will normally be required which will vary with the type of roof and the local context.

### *9.2.3.6 Green walls*

*Green walls are systems that encourage vegetation growth up and across walls*

There is scope for improving the appearance of some parts of industrial areas (boundary walls for example), using simple green walls, ideally with drainage modifications to replace a section of below-ground pipe and with a gravel filter drain connected to underlying and surrounding soil, in which climbing plants are rooted.

**Figure 9.7** Green walls rooted in gravel filter drains, Barcelona, Spain.

#### 9.2.3.6.1 Advantages

Green walls can provide excellent opportunities to reduce ambient air temperature in hotter climates, especially in heavily urbanised settings and will improve air quality and create useful habitat. Their appearance can also significantly enhance the appearance of a building or wider location. Green walls have some potential to reduce runoff *if designed for that purpose* with minimal land take.

#### 9.2.3.6.2 Disadvantages

Rarely designed as SUDS features for managing stormwater, green walls can also demand more maintenance to ensure their function, appearance and safety for observers and the local community (design considerations discussed in Dunnet and Kingsbury (2008)).

### *9.2.3.7 Pervious pavements*

*Load bearing pavements suitable for vehicles which allow rainwater to infiltrate through the surface to underlying structures for flow attenuation and, where appropriate, infiltration*

- Two types of surface: porous surfaces which allow infiltration across the whole surface (e.g., porous asphalt or porous concrete), and permeable surfaces comprising impervious material which allows movement of water through it by a pattern of voids (e.g., concrete block paving) (Wilson *et al.* 2004).
- Various commercial materials are available to create pervious surfaces in addition to the above examples, including hybrid soft/hard engineering surfaces which utilise grass within a supporting matrix of concrete or plastic. The latter are widely used, but are not suited to regular, frequent dense coverage by parked vehicles, which shade out the vegetation.
- Technical details covering a broad variety of types of pavement and sub-base, including materials and structural factors are given in CIRIA (2015).

**Figure 9.8** Above: Permeable blacktop serving the car park for a meat trading market near Dingwall Scotland, UK. Below: Hybrid concrete/grass permeable surface, with exceedance flow collection drain, serving a coach and car parking area at a water utility installation in South Korea.

#### 9.2.3.7.1 Advantages

Permeable pavements can provide flow attenuation and diffuse pollutant degradation at zero land take, for example on a car park for employees/visitors/customers.

#### 9.2.3.7.2 Disadvantages

There are specific maintenance requirements, which if not respected will result in shortened life of the features. Cost per unit area is generally greater than equivalent hard surfaces (if no gullies and fewer pipes are needed then cost difference can be offset). Generally not regarded as suitable for heavy vehicles in the UK.

### *9.2.3.8 Filter drains*

*Linear drainage features comprising a trench filled with a permeable aggregate and often including a perforated pipe in the base of the trench*

- Shallow trenches of 1–2 m depth (min. 0.5 m).
- Treatment enhanced if runoff first passes over a grass filter strip.
- Void space in stone fill for flow attenuation can be optimised by a downstream flow control.
- If a length of drain exceeds 10 m, it needs to be served by an inspection sump/s (catchpits); the maximum interval between such points may be 90 m (CIRIA, 2015).

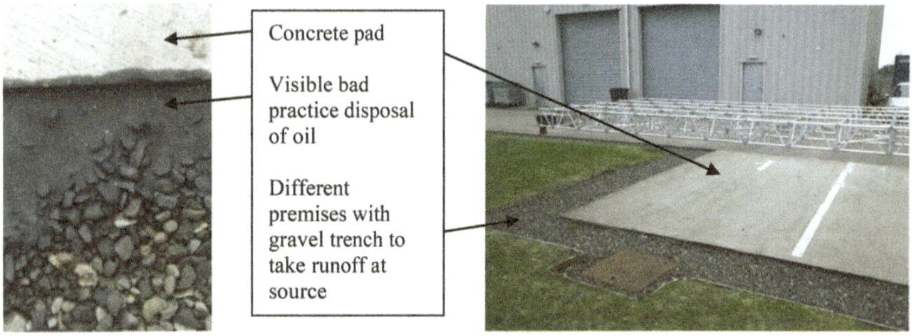


**Figure 9.9** *Left*: Close up of a gravel drain revealing ease of seeing poor practice oil disposal. *Right*: Filter drains at edge of concrete pad, taking runoff from property in industrial estate, Kinross, Scotland, UK.

#### 9.2.3.8.1 Advantages

Limited land take and can provide initial flow attenuation and diffuse pollutant degradation at source, and tolerate occasional intrusion by vehicles on site. If damaged (e.g., compacted or silted) relatively straightforward to repair.

#### 9.2.3.8.2 Disadvantages

Exposure to subsoil from construction work, or any of various possible organic materials likely to be on an industrial yard (waste paper and rags, spent grain, draff, building sand, animal feed spills, reject paper fibre, or viscous materials, etc.) will shorten the functional life of the features. Clean up costs can be avoided by tighter control over such materials on site.

### *9.2.3.9 Extended detention basins*

*Vegetated, landscaped depressions which temporarily store water to attenuate runoff. Small areas of permanent pools may be features in an enhanced vegetated basin.*

- Basins are relatively shallow to allow grasses and other vegetation to survive intermittent inundation, base topography needs to allow maximum area to flood, but excessively uniform grading is costly, reduces amenity for employees and visitors (looks un-natural and is poor for wildlife).
- Flow controlled outlets are essential (e.g., weirs or vertical risers, with overflow for larger storms).
- Excessive (non-functional) depths of a basin feature are consequences of conventional piped drainage inflows, where the basin is an end-of-pipe feature (avoid if possible).
- Sedimentation forebay area close to inlet facilitates maintenance.
- Final pool or wetland enhances amenity and improves water quality (prevents mobilisation of sediment during drain-down, Schueler *et al.* (1992)).
- On permeable soils in industrial premises, or contaminated land, a liner is needed to protect groundwater (see hotspots, section 4.2).
- Access for sediment removal, and for dealing with pollution incidents or fire-water is vital.
- Length to width ratio 3:1 to 5:1, side slopes <1 in 3, water depth of flow <100 mm, flow velocity not greater than 0.3 m/s and minimum travel time from inlet to outlet should be at least 9 minutes (CIRIA, 2015).

**Figure 9.10** Extended detention basin, with source control feeder swale serving service road, Kinross, UK.

#### 9.2.3.9.1 Advantages

Vegetation in detention basins is exposed to sunlight (not permanently covered by a metre or more of turbid water) which favours pollutant breakdown, and soils, at least in part, will be aerobic treatment zones for hydrocarbon contamination (Napier *et al.* 2009). Oil spills or other visible pollutants will be obvious to management and can be addressed and future incidents prevented. Good 'catch-all' site features.

#### 9.2.3.9.2 Disadvantages

Land take is significant. Schueler (1992) notes that detention basins are among the least expensive features to create, but can be the most expensive to maintain.

### *9.2.3.10 Retention ponds and stormwater wetlands*

*Features designed to temporarily store water above permanently wet pools sized to allow sedimentation and biological removal of pollutants*

- Irregular shapes enhance amenity, and modular design allows better fit with available space.
- Shallow peripheral zone of emergent vegetation deters casual access to deeper water (1–2 m).
- Inlet should be designed to facilitate removal of oil and other pollutants behind an oil boom or baffle, and for pumping fire-water if necessary.
- Pond should be located a safe distance from any high risks areas, to allow access throughout a pollution crisis, but ideally overlooked and served by good emergency vehicle and pedestrian access (meeting incident response needs, and day-to-day relaxation for employees/contractors).
- A sediment forebay is an essential feature for long term maintenance.

**Figure 9.11** Retention pond in modular arrangement (3 ponds and a wetland) receiving runoff from a timber industries business park in Scotland, UK. Note the provision for management of incidents as well as day to day runoff. Swales or filter drains provide the source controls around the site.

The differences between the various types of ponds, and a rationale for the nomenclature is given in Campbell *et al.* (2004).

#### 9.2.3.10.1 Advantages

Almost guarantees river quality runoff from a commercial/industrial catchment. A source of fire water, easily aligned with flood risk management due to large surface area. Provides excellent opportunities for major incident management. Opportunities for modular design to fit available space and landscapes (see Figure 9.11).

#### 9.2.3.10.2 Disadvantages

Land take is a significant consideration. Maintenance is modest until sediment removal is required in the sediment forebay at least, when significant costs can be anticipated. Safety concerns.

## 9.3 OTHER DRAINAGE FEATURES

### 9.3.1 Inspection chambers

Where buried pipe systems are present, a final inspection chamber should be provided at each industrial site just within the premises, and prior to connection of the site stormwater drainage to a public or other network or watercourse. If large enough, with easier access than a conventional manhole, such a chamber enables two pollution prevention activities:

(i) Site management of day to day pollution risks and performance of staff and facilities.
(ii) Access to a drainage point at maximum distance from the factory or warehouses or other storage areas, so that fire-water can be pumped out safely in the event of a catastrophic fire, or pollutants recovered from a major oil or other pollutant leak or accident.

For (i), an oil absorbent pad or cushion can be suspended in the drainage chamber and regularly inspected for evidence of oil contamination (and replaced periodically) as part of a site foreman's check on environmental and safety issues. In the event of leaks (not always evident at the installations involved, for example corrosion inside a steam heated oil tank whereby oil enters the steam line and exits to drain, or a crack in a section of partially buried feedline from a storage tank to a boiler or production unit), an oil absorbent cushion or net bag of absorbent material on a rope in the inspection chamber will allow management intervention before too much valuable product is lost.

For (ii) a proper inspection chamber is important for the management of major incidents involving chemicals or oil spills, or fires when pesticides or other toxic contaminants (sometimes derived from material arising from combustion of the fabric of the buildings) is mobilised in firewater runoff. The ability to close the outlet in an inspection chamber on a stormwater drain allows capture of polluting material to be thenmanaged according to its nature (as set out in a company policy and contingency plan).

Simple infrastructure features such as this would go some way to justify the description 'industrial estate'.

### 9.3.2 Silt traps and oil interceptors

These features are standard for heavily contaminated areas, to capture gross contamination and for that purpose are useful if adequately maintained. Close to source, they allow for recovery of resource (product or raw material) before it becomes too contaminated by other material. Sealed sumps, located inside factory units are a preferred option. Taken collectively across multiple premises, silt traps and oil interceptors taking contaminated runoff are not capable of producing river quality outflows, which is the aim of SUDS features for runoff from the developed land in a drainage catchment. Horner *et al.* (1994) note that for runoff, vegetated systems perform better than oil interceptors, the latter being incapable of reducing oil concentrations to less than 10 mg/l. Oil interceptors are pre-treatment measures for higher risk areas (Schueler *et al.* 1992). Ideally they discharge to a foul sewer for full treatment.

### 9.3.3 Packaged filtration units and vortex separators

For retrofit situations, the application of several of the techniques set out in section 3.2 is impractical, due to space constraints. Those techniques are most suited to green-field developments, or some redevelopment opportunities. That does not mean that there are no treatment options where chronic industrial estate or factory pollution due to contaminated stormwater is known to be an issue. In addition to filter drains and bioretention units, filter trenches and partial treatment structures are available from specialist suppliers for industrial/commercial yard areas (CIRIA, 2015). There are many proprietary treatment units now available for retrofit in conventional piped drainage systems, produced by a number of specialist stormwater-focused companies. If fitted within individual industrial premises such devices meet the polluter pays principle. Maintenance must be undertaken regularly for effective sustained performance, including sediment removal and periodic replacement of filter bags or media, depending on the degree of contamination. In addition, vortex sediment traps and some filtration units may also be capable of providing a downstream pollutant removal capability, since they have been designed for stormwater sewer systems. That allows for a treatment train retrofit programme by public utilities and agencies to restore the quality of intensively industrialised watercourses in towns and cities.

## 9.4 INFRASTRUCTURE STRATEGY

### 9.4.1 Water quality and pollution risks for industrial premises and estates

Designing the drainage infrastructure for an industrial estate requires consideration of the two primary categories of pollution risk identified in the opening section of this chapter: incident risk management and contingency planning, and designing to capture and treat diffuse pollutants in runoff.

Details of pollution risk have been described elsewhere (e.g., Harper, 1987; D'Arcy & Bayes, 1995). The diffuse nature of industrial estate pollution, with multiple sources, requires a best management practice approach, involving good housekeeping measures such as bunding oil and chemical tanks as well as provision of drainage infrastructure (SUDS features) to capture pollutants entrained in runoff (Campbell *et al.* 2004). An open drainage structure has great merit since it allows for easy inspection of drainage by management as well as statutory agencies, and allows for quick and effective damage-limitation responses in the event of traffic or other actions. Hydrocarbons are generally the contaminants most likely to exceed recommended limits in urban stream sediments (e.g., Wilson *et al.* 2005). Various studies have independently demonstrated that soft engineering SUDS favour not just capture of diffuse pollutants, but provide degradation *in situ* (Horner *et al.* 1994; Napier *et al.* 2009; Leroy, 2016). For hydrocarbons, that degradation reduces the cost of disposal when the accumulated sediment is eventually removed from the features (avoids likelihood of need for expensive disposal at an approved landfill site). Table 9.2 summarises how different drainage features can be important elements of a strategic approach to infrastructure provision when planning industrial and commercial developments.

A third category, usually addressed by guidance and best practice recommendations for building site construction, rather than the infrastructure itself, is the misconnection of process effluents into surface water drains. Typical errors include sewage as well as various wash waters and other effluents. Causes of pollution vary from structural errors in construction, to *ad hoc* steam cleaning of plant or vehicles. A shared steam cleaning area on an industrial estate would be an infrastructure option which could be a good way to ensure proper collection and treatment, or recycling, of wash waters, if designed and agreed in consultation with the local water utility for sewer connection. Use of mobile plant for irregular steam cleaning may still occur however, and in a conventional industrial estate served by a separate sewer system will result in pollution. A pipe free system, using SUDS techniques, should reduce risks of pollution by operators unaware of the different types of sewer on conventional premises. That is as important for commercial as for industrial premises.

### 9.4.2 Spatial application of SUDS

This refers to ***where*** the technology should be applied across a development. It is concerned with responsibilities, with ownership of sources of pollution and

**Table 9.2** Usefulness (Good, Medium, Limited) of various types of drainage features for industrial/commercial premises and estates.

| Feature | Contingency Plans for Major Incidents | Diffuse Pollutant Capture* | Comments:<br>1. Pollutant degradation in situ<br>2. Storage capacity<br>3. High visibility drives cleaner practice |
|---|---|---|---|
| Grass filter strip | Limited | Good | **1** if not overloaded, **3** |
| Grass swale | Good | Medium | **1,2,3** Large volume for incident management |
| Under-drained swale | Medium/Good | Medium | **1,2,3** Less storage capacity at surface than a swale |
| Permeable pavement | Limited/ Medium | Good | **1,2** May be difficult to access contaminant in event of an accident or spill, storage dependent on sub-base void space volume |
| Gravel filter drain | Medium | Medium | **1,3** Limited storage, but slows release from source area if via a flow control structure |
| Bioretention units | Limited | Medium | **1,3** Visibility of bad practice disposal |
| Green roofs | Limited | Medium | For air pollution (e.g., factory unit air vent release) |
| Green walls | Limited | Limited | **1**? Green walls rooted in a filter drain may be good? |
| Detention basin | Good | Medium/Good | **1,2** Drain-down with facility to close outlet |
| Retention pond | Good | Good | **1,2** Quiescent water body for skimming oil after spill, long residence time. Oil boom deployment easy. Good for 'polishing' drainage quality routinely. Fire-water source. |
| Oil interceptors | Limited | Limited | Close to high-risk installations, 'pre-treatment for fuelling points, ideally outflow to foul sewer |
| Silt traps | Medium | Limited | If not too close to at-risk installations, may be useful as pumping sumps for fire water |
| Inspection chambers | Medium/Good | Limited | A large chamber prior to drainage connection to public network is good for controlling outflow of drainage and fire-water, or pumping latter for safe disposal. |
| High intensity filtration units | Limited | Retrofit option | Provides ways to address existing chronic pollution |
| Vortex separators | Limited | Retrofit option | Provides ways to address existing chronic pollution |

*Treatment effectiveness – see International Stormwater BMP Database, [2016] accessed at www.bmpdatabase.org, and CIRIA (2015).

stormwater runoff, with contingency planning and risk management. It allows for policies by planning authorities and environmental regulators to require SUDS as source controls, for conveyance and for end-of-system.

#### 9.4.2.1 At source and on site

It is necessary to have source control SUDS at each unit on an industrial estate, to allow capture of pollutants as close to source as possible, and within the remit of site management, for each of the premises on the estate. This allows for management at each site to see and act upon incidents, as well as ensure maintenance is undertaken for the site's own drainage infrastructure. It also establishes responsibility for each site owner/operator (consistent with the *polluter pays principle*). Public sector (or private estate managers) should only be responsible for off-site drainage, including SUDS features.

All premises are not alike and the Maryland stormwater manual (Maryland, 2009) identified stormwater hotspots, defined as

> 'a land-use or activity that generates higher concentrations of hydrocarbons, trace metals or toxicants than are found in typical stormwater runoff, based on monitoring studies.'

Table 9.3, modified from the Maryland guidance identifies some hotspots likely to occur on industrial/commercial premises. To protect groundwater resources, runoff from hotspots cannot be allowed to infiltrate, and a greater level of stormwater treatment is required at hotspot sites. Hotspots are contrasted with lower risk urban areas such as residential streets and rural highways, residential properties, office developments, and non-industrial rooftops.

**Table 9.3** Some identified stormwater hotspots on industrial/commercial premises.

| Hotspot | Comments |
|---|---|
| Outdoor storage of scrap materials | Risks of oil, PCBs, PAHs, toxic metals, and other pollutants |
| Recycling facilities | Depends on materials on site |
| Vehicle fuelling stations | Oil interceptors |
| Vehicle service & maintenance facilities | Sealed sumps, indoor activities preferable |
| Vehicle & equipment cleaning facilities | Oil-detergent mixes, solvents |
| Fleet storage areas (bus, truck etc.) | Hydrocarbons |
| Commercial parking lots | Hydrocarbons |
| Industrial sites with outdoor storage of potential pollutants | Raw materials, products and by-products, wastes |
| Outdoor loading/unloading areas | Solids and liquids, foods and other degradable materials, not just toxic substances |
| Council depots | Local authority road sweepers, refuse trucks, fuel, maintenance work & cleaning |
| Marinas and hull paining/ refurbishment areas | Detergent, anti-fouling paints, fuelling points |

*Source:* Aadapted from Maryland, 2009.

#### *9.4.2.2 Conveyance swales or drains*

The range of pollutants being transported in and around industrial estates also warrants use of open, passive treatment features such as grass swales or filter drains for conveyance to a final SUDS feature. Typically, the conveyance features for the drainage will be alongside the access roads in an estate. A road tanker accident could occur away from the destination premises, and a roadside intervention feature such as a swale or filter drain allows for local pollution prevention action to isolate and contain spills. The technology will also provide a secondary treatment function after the source/site controls. Swales also allow visible pollution incidents such as oil or dyes, or toxic effects on vegetation, to be easily seen and hence dealt with.

#### *9.4.2.3 Regional (whole estate) features*

It is impractical, unjustifiable, and probably undesirable to require a stormwater wetland or retention pond for every factory or other premises on an industrial estate. In normal circumstances a final retention pond, if protected from excessive loading by source controls and road side swales as above, would produce river quality drainage even from an industrial estate (D'Arcy *et al.* 2007). And if and when a major incident occurred, the end-of-system retention pond or wetland would also allow management of a catastrophic pollution accident (e.g., a major sludge or oil leak).

Considering an estate or business park as a whole, the above application of the technology within and across the estate constitutes a 'treatment train' of successive stormwater management measures. In practice a 'treatment web' may be a more appropriate description of the drainage features; some in series, but not all in a single strand.

### 9.4.3 Treatment capability

For an entire estate, or an extensive factory such as an automobile assembly plant, cable manufacture, or a brewery, this is a function of two different considerations:

(a) the collective master plan overview of the treatment impact of an overall scheme, which aims to produce river quality drainage from the development as a whole
(b) the ***type*** of treatment processes that are present in any single SUDS component of the development infrastructure.

The first category (a) is the sum of the second, (b). Treatment capability is concerned with how a feature will capture pollutants entrained in stormwater runoff, and also the fate of pollutants in the SUDS features. It is also important to consider what proportion of the pollutant load in the runoff will be treated by the feature, and how much by-passes it, if any? The essence of a best practice approach is that environmental protection should be achieved if units with known and demonstrable performance are designed on the basis of the pollutants likely to be present in runoff from the catchment, and are constructed and installed correctly, and maintained.

Pollutant capture may be by sedimentation, filtration, adsorption, or storage. Biodegradation and decomposition may be achieved by ultraviolet light and microbes, and pollutant uptake can occur in vegetation and, for some substances, in the food chain. Treatment processes and factors affecting the removal of various pollutants have been considered in various studies and guidance publications, for example, Horner *et al.* (1994), Campbell *et al.* (2004), Ellis *et al.* (2013), and CIRIA (2015).

The fate of pollutants may involve biodegradation in the water column (e.g., in a retention pond), or in aerobic soils in a swale or detention basin, or on microbial films on stone media in permeable pavements and filter drains, as well as on stems of higher plants. Degradation may occur on the surface of vegetation exposed to ultraviolet light, or on hard surfaces such as permeable pavement (less likely to be washed off by sheet flow into drainage systems than conventional pavement). The treatment capabilities and fate of pollutants are important for maintenance considerations and disposal costs of accumulated material in SUDS, as are sediment storage capacities within a SUDS feature (Napier *et al.* 2009).

A *treatment train* of measures allows for sequential capture of pollutants in successive features across a development; sometimes within premises, certainly from the source control at each of the premises, through to a regional pond or wetland. The different pollutant capture and/or degradation properties of various types of features can be considered in selecting features for use in such a treatment train. The cumulative removal of pollutants from runoff across a development will enhance the biodiversity and amenity value of any end-of-system feature such as a pond or wetland.

A *stormwater management train*, by contrast seeks to include all aspects of stormwater management needs; flood risk as well as pollution risk. It may seek merely flow attenuation at source for roof runoff, some treatment for driveway runoff, as well as provision of treatment for street runoff and/or perhaps end-of-pipe features providing flood risk management functions too, with due allowance for exceedence flows to move across a development without damage. It incorporates the aspirations of the treatment train in a broader context, rather than simply for example allowing high flows of possibly cleaner runoff to by-pass treatment features, but possibly still present a local flooding risk.

## 9.5 DISCUSSION

### 9.5.1 Treatment trains

The grass swales in Figure 9.3 and gravel drains in Figure 9.9, both provide an open drainage system in which any bad practices such as disposal of waste oil, or accidental spills, can readily be seen, for example by management, and questions asked and remedial actions taken, with prevention in future. Such features therefore could lead to *reductions in loading*, as well as provide a first stage, close to source, interception and treatment capability.

### 9.5.2 Compromise or lateral thinking

If local circumstances on a site preclude a technique which was originally favoured, then rather than compromise on the design specifications, it is generally preferable to select a different technique.

Selection of one type of drainage option may compromise another downstream, for example if a pre-treatment feature requiring drainage at depth is provided, based on a preference for conventional piped drainage, it will be difficult to provide a naturally shallow surface feature such as a swale without massively expanding land take for a deep steep-sided channel into which the conventional deep drains can connect. Making the sides steeper to reduce land take may create hazards. Putting a shallow, source control, swale on top of the deeper drain might be a better option, or using intermittent bioretention planter units.

### 9.5.3 Technology development for managing design and construction risks

How to address uncertainty in design and construction, especially where supervision may be limited? There may be standardisation benefits in 'packaging' for example bioretention features into box structures that can be selected by developers from a look-up table and applied according to unit roof/driveway/yard area for plot by plot applications (Campbell & D'Arcy, 2016). For source control units, there is considerable scope for off-the-shelf SUDS features which have proven performance in relation to runoff from specified catchment areas.

### 9.5.4 Catchment initiatives and retrofits

Prior to understanding diffuse pollution and its impact on baseline water quality in an urban/industrialised catchment, even environmental regulators were often inclined to dismiss minor sources on an individual site. A common error is comparing runoff to, for example, a process effluent or a chemical spill, instead of comparing runoff quality with river quality. Catchment scale engagement initiatives are often needed to highlight the collective impact of many premises, at each of which only minor contamination may have been recognised and ignored in isolation from wider perspectives.

## 9.6 CONCLUSIONS

There is a broad international consensus now on best management practices to capture diffuse source pollutants and these play a major role in stormwater management for wider purposes too. The successful implementation of measures depends on adoption of a clear strategy, with clear objectives that can be set out in early consultations between architects, developers, planners and environmental regulators, whether for new build or retrofit.

All new and redeveloped industrial estates and premises should be planned and designed with provision for the following:

- Bund walls and other housekeeping measures at oil, chemical and other potential pollutant storage installations, and canopies over loading-unloading areas to exclude rainfall runoff and allow recovery of spilled material in sealed sumps.
- A location at the rear of premises, for connecting potentially polluting drainage to foul sewer (subject to prior agreement of the drainage authority). Runoff from vehicle wash-bays, refuelling areas, oil and chemical storage areas and areas where potentially polluting material is handled should be directed to sewer after pre-agreement.

For drainage infrastructure:

- SUDS source control features at each of the premises on an industrial estate.
- Open drainage systems such as swales for conveyance of runoff from each of the premises to downstream control features.
- A retention pond or stormwater wetland should be the end-of-system final control, prior to discharge to watercourse. A modular arrangement has benefits for incident management.
- Features should be designed with contingency planning in mind too, as part of accident management (including oil, chemical and other spills and leaks, and fires).
- An inspection chamber, including provision to isolate outflow, should be the final part of a drainage system on each of the premises on an estate, if it is to be described as an *industrial* or *commercial* estate.
- Silt traps and/or oil interceptors should be provided where occupancy by a business with high-risk characteristics requires that.
- Retrofit programmes in dense developments should not ignore the potential benefits of commercial stormwater treatment technology, especially on a site-by-site basis (polluter pays principle).

An integrated, multiple benefits philosophy is essential to maximise benefits for industry and commerce, for developers and local agencies, and for the environment. If provision is made for landscape quality, access or overview by staff and visitors, and habitat creation – the broad span of benefits from multifunctional landscape features implicit in the SUDS concept – the infrastructure will contribute to well-being and creating attractive places to work, as well as drainage and pollution prevention functions. Last but not least, the potential well-being benefits of green infrastructure for management, employees and contractors on industrial and commercial premises should not be overlooked.

## 9.7 ACKNOWLEDGEMENTS

The authors are grateful to Neil Campbell and Andrew Hemingway for comments on drafts of this paper. All photographs are copyright Brian J. D'Arcy, unless otherwise credited.

## REFERENCES

Arias L., Grimade J.-C. and Bertrand-Krajewski J.-L. (2016). First Results of Hydrological Performances of Three Different Green Roofs. Novatech Abstracts, Lyon, France, p. 77.

Boogaard F. C., Lucke T., Sommer H., De Beer J., van de Giesen N. C. and van de Ven F. H. M. (2014). Lessons Learned from Over Two Decades of Global Swale Use. 13th International Conference on Urban Drainage, 7–12 September 2014, Sarawak, Malaysia.

Campbell N., D'Arcy B., Frost A., Novotny V. and Sansom A. (2004). Diffuse Pollution: An Introduction to the Problems and Solutions. IWA Publishing, London. ISBN: 1 900222 53 1.

Campbell N. S. and D'Arcy B. J. (2016). Controlling uncertainties – SUDS in a box? *Blog*, http://www.enviroexperience.co.uk

Casal-Campos C., Jefferies C. and Perales Momparler S. (2012). Selecting SUDS in the Valencia Region of Spain. Published March 2012, 7 (1) wpt2012001, DOI: 10.2166/wpt.2012.001.

CIRIA (2015). The SuDS Manual. CIRIA, London C753, 968pp. www.ciria.org, ISBN: 978-0-86017-760-9.

D'Arcy B. J. and Bayes C. D. (1995). Industrial Estates: A Problem. In Pratt C. (ed.), Proceedings of the Tenth Meeting of the Standing Conference on Stormwater Source Control. School of The Built Environment, Coventry University, Coventry. ISBN 0 905949 33 1.

D'Arcy B. J., Rosenqvist T., Mitchell G., Kellagher R. and Billett S. (2007). Restoration challenges for urban watercourses. *Water Science & Technology*, **55**(3), 1–7.

Dunnett N. and Kingsbury N. (2008). Planting Green Roofs and Living Walls. Timber Press, Portland Oregon, and London UK. ISBN 13: 978-0-88192-911-9.

Ellis J. B., Deutsch J.-C., Mouchel J.-M., Scholes L. and Revitt D. M. (2004). Multicriteria decision approaches to support sustainable drainage options for the treatment of highway and urban runoff. *Science of the Total Environment* **334–335**, 251–260.

Ellis J. B., Revitt M. and Lundy L. (2013). Behaviour of Selected Emerging Pollutants in Stormwater Best Management Practices (BMPs). Novatch, Lyon, France.

Harper E. (1987). Environmental Impact – the role of the water authorities in the UK. In: Fires Involving Pesticides – The Rhine Incident, Proceedings of a Seminar in York, 25/26 September 1987, British Agrochemicals Association, Peterborough, UK, pp. 36–45.

Heal K., Scholz M., Wilby N. and Homer B. (2005). The Caw Burn SUDS: Performance of a Settlement Pond/Wetland SUDS Retrofit. Proceedings of Third National Conference on Sustainable Drainage, 20–21 June 2005, Coventry University. ISBN: 1 84600 007 6.

Horner R. R., Skupien J. J., Livingston E. H. and Shaver H. E. (1994). Fundamentals of Urban Runoff Management: Technical and Institutional Issues. Terence Institute, Washington DC.

Hreggviosdottir H. and Arnadottir H. (2016). Urridaholt neighbourhood Iceland – Sustainable drainage solutions interwoven in the urban pattern – a co-operative approach.

Ibanez M. (2015). Characterisation of Water Pollution Problems and Solutions from Industrial Estates – East Tullos burn, MSc dissertation, University of Abertay, Aberdeen (Scotland), September 2015, Dundee, UK.

International Stormwater BMP Database (2016). www.bmpdatabase.org (accessed 8 August 2016)

Leroy M.-C., Marcotte S., Le Derf F., Legras M., Moncond'huy V. and Portet-Koltalo (2016). Runoff Treatment by Roadside Swales: Soil and Plants Improvements. Novatech Abstracts, Lyon, France, p. 223.

Napier F., Jefferies C., Heal K. V., Fogg P., D'Arcy B. J. and Clarke R. (2009). Evidence of traffic-related pollutant control in soil-based Sustainable Urban Drainage Systems (SUDS). *Water Science and Technology*, **60**(1), 221–30.

Porter E. (1973). Pollution in Four Industrialised Estuaries. The RCoEP, HMSO, London.

Schueler T. R., Kumble P. A. and Heraty M. A. (1992). A Current Assessment of Best Management Practices. Metropolitan Council of Governments, Washington DC.

Stahre P. (2006). Sustainability in Urban Storm Drainage, Planning and Examples. Svenskt Vatten, Malmo, Sweden. ISBN: 91-85159-20-4.

UNEP (1997). The Environmental Management of Industrial Estates, UNEP IE Technical Report No. 39, United Nations Environment Programme, Industry and Environment. ISBN: 92-807-1652-2.

# Chapter 10

# Low impact development features: hydrological and environmental effects

*Marla C. Maniquiz-Redillas and Lee-Hyung Kim*

*Department of Civil and Environmental Engineering, Kongju National University, Cheonan, Chungnamdo, Korea (Email: marlcmani@kongju.ac.kr, leehyung@kongju.ac.kr)*

## 10.1 INTRODUCTION

Roads, parking lots and other transport-related impervious surfaces are both hydrological and ecological disturbances. Stormwater discharges from these land uses can cause adverse effects and lead to significant degradation of environmental values of the urban aquatic ecosystems into which they discharge. The effects can be significant from physical changes in streams due to hydrological impacts, through to deposition of contaminants into ecosystems or groundwater. The impacts have historically been more easily seen in heavily industrialized countries, where large rivers have been severely impacted.

The importance of stormwater management in industrial sites is recognized worldwide since the range and quantity of contaminants present in discharged stormwater is typically higher than other land uses. Typical substances present in industrial sites include acids, alkalis, degradable organic residues, detergents, disinfectants, dyes, engine coolants, fertilizers, fuel, lubricants, metal solutions, pharmaceuticals, salts, poisons and solvents. Stormwater runoff from roofs, parking lots, driveways, exterior materials storage and process areas should be effectively managed to avoid flooding and contamination risks especially when stored toxic chemicals are flushed off-site or washed off during rainfall events and discharged to receiving waterbodies. All risks should be considered, from the

© IWA Publishing 2017. Wealth Creation without Pollution: Designing for Industry, Ecobusiness Parks and Industrial Estates
Brian D'Arcy, Lee-Hyung Kim, Marla Maniquiz-Redillas
doi: 10.2166/9781780408330_135

individually minor ones, to a major disaster involving harm to people, property and/or ecosystems. A small or repeated chemical discharge over an extended period may lead to contaminant accumulation in an aquifer, sediment or confined surface waters. Even chemical spills (if poorly managed) can overwhelm the capacity of receiving water resources to assimilate or break down contaminants and result in degraded water values.

The standard management methods at industrial sites including environmental management plans, spill procedures and treatment devices are necessary approaches. However, the integration of low impact development (LID) principles into the site design process can achieve a sustainable stormwater solution and provide better environmental outcomes – both for the receiving water habitats and for the urban environment (Table 10.1). It provides an opportunity for innovation both in the site design and in the use of stormwater treatment devices. While larger existing industrial areas or industrial activities are priorities for management, future developments will certainly take place for industrial purposes and can add LID to their management of risks. These new industrial sites should be designed with LID principles to address issues of physical and biochemical impacts of catchment urbanization on the aquatic ecosystem. Many factors influence aquatic ecosystem health but often changes in catchment hydrology and poor stormwater quality are the main driving factors in ecosystem health deterioration.

**Table 10.1** Comparison of the characteristics of stormwater in residential and industrial areas.

| Attribute | Residential | Industrial |
|---|---|---|
| Major contaminants | • Particulates, nutrients, organics | • Particulates, nutrients, organics, heavy metals, hydrocarbons, and other toxic materials (acids, alkali, chlorine, etc.) |
| Sources | • Gardens, lawns, roofs, driveways | • Parking lots, roofs, roadways, etc. |
| Stormwater control | • Site/source control (small scale LID) | • Site/source control (small scale LID)<br>• Regional control (large scale LID) |
| Design considerations | • Small storage capacity (small runoff coefficient)<br>• Sand could provide high infiltration rate as well as filtration of contaminants | • Big storage capacity (high runoff coefficient)<br>• Filter media with high adsorption capacity could provide high infiltration rate as well as filtration of highly polluted runoff<br>• Groundwater quality management should be considered |

Commercial areas and industrial complexes are places where various places of business are located and where there is high traffic of people and vehicles. Of the two, commercial areas lack a wide area of space, thus small-scale decentralized type facilities are established. However, in industrial complexes, LID can be installed in connection with detention facilities to prepare for the possibility of toxic chemical outflow (Figure 10.1).

(**a**) Commercial area (California, USA)

(**b**) Industrial area (Ohio, USA)

**Figure 10.1** LID application examples in commercial and industrial areas.

In this chapter, the importance of stormwater management measures in industrial catchments is addressed. Particularly, placing an emphasis on the considerations and guidelines in designing and sizing LID practices and techniques to address issues of physical and biochemical impacts of urbanization in industrial sites on the aquatic ecosystem. The performance, hydrological and environmental effects of LID are presented. It is hoped that the information provided here could offer guidance for both new development and redevelopment of industrial sites and contribute to the integral change in stormwater management in industrial sites, that is to produce smaller less obtrusive facilities that are more aesthetic and less burdensome on those responsible for their long-term maintenance and performance while sustaining a healthy environment.

## 10.2 LOW IMPACT DEVELOPMENT (LID)

### 10.2.1 Principles and goals

LID is a stormwater management approach most influentially used in Prince Georges County, Maryland, United States in the early 1990s and also commonly used in North America, New Zealand and South Korea. LID complements other urban planning techniques such as 'Sustainable Urban Drainage Systems (SUDS)' in the UK, 'Water Sensitive Urban Design (WSUD)' in Australia, 'Alternative Techniques (AT)' in French-speaking countries, 'Stormwater Control Measures (SCMs),' 'Best Management Practices (BMPs),' 'Smart Growth,' 'Green Building,' 'Sustainable Development,' and 'Green Infrastructure (GI)' in the United States (Fletcher *et al.* 2015). These stormwater management strategies attempt to restore the natural water cycle and use design practices and techniques that effectively capture, infiltrate, filter, store, evaporate, and detain runoff close to its source (Prince George's County [PGC], 1999). Rather than conveying the runoff from small frequent storm events directly into underground pipes and drainage systems for discharge offsite, LID practices dissipate and infiltrate stormwater runoff with landscape features and, where practical, permeable surfaces located onsite, thereby reducing runoff volumes and filtering runoff before it leaves the site. LID design techniques and practices need to look at the major development features of a project, including project green space areas and landscaping, streetscapes, parking lots, sidewalks, and medians. LID is a versatile approach that can be applied to new development, urban retrofits, redevelopment, capital improvement and revitalization projects (Coffman, 2002; Ferguson, 2002; Lloyd *et al.* 2002; US Department of Defense [DoD], 2004; Davis, 2005). In addition, the use of LID practices is less costly to implement and generally more aesthetically pleasing than traditional conveyance systems, and integrates well into the existing infrastructure (US Environmental Protection Agency [EPA], 2000).

LID's approach to urban planning and design aims to minimize the hydrological impacts of urban development on the surrounding environment (Table 10.2). Both stormwater management and LID are directed at providing flood control, flow management, and water quality improvements (Davis, 2005; Mull, 2005; County of San Diego [CSD], 2007). The goal of LID site design is to reduce the generation of stormwater runoff and to treat pollutant loads where they are generated. This is accomplished first with appropriate site planning and then by directing stormwater towards small-scale systems that are dispersed throughout the site with the purpose of managing water in an evenly distributed manner. These distributed systems allow for downsizing or elimination of stormwater ponds, curbs, and gutters. Designers and developers can select the LID technologies relevant to the site's topographic and climatic conditions that are appropriate to meet stormwater control requirements and specific project constraints and opportunities (CSD, 2007; Guillette, 2007).

**Table 10.2** Strategies for LID approach.

| Strategy | Content |
|---|---|
| Preserve and regenerate vegetation and soil | • Preserving vegetation on the original soil as much as possible while maintaining the natural drainage pattern, topography, sinking place, etc.<br>• Selecting types of plants and planting them again to recover native vegetation<br>• Preserving hydrological soil type A and B (moderate to high permeability) with good drainage as much as possible<br>• Minimizing soil consolidation and disturbance<br>• Using fertilizers to recover the health of soil consolidated by construction<br>• Maintaining natural drainage system and topography of property and considering them in property design<br>• Reorganizing property to delay surface runoff and increase infiltration function using the existing topography of the property<br>• Using open vegetated swale and natural vegetation drainage pattern for extension of water ways and flow distribution<br>• Blocking the flow by using detention systems of nature such as vegetated swale or rain garden<br>• Minimizing curbs or bypass drainage systems and designing so that impervious surface areas are not continued<br>• Extending rainwater's detention time in the sources by connecting impervious areas and vegetated areas<br>• Maintaining low slope |
| Minimize impervious areas | • Need common value or consensus based on co-operation of property designers, planners, engineers, landscape planners and architects<br>• Maintaining land cover type, imperviousness rate, connectivity, hydrological soil type, natural flow pattern and retention characteristics<br>• Minimizing disturbances, preserving open vegetated swale and soil with high infiltration rate, establishing nonpoint source treatment facilities in soil with high infiltration rate, etc.<br>• Minimizing impervious surface areas such as roofs, roads and parking lots<br>• Installing impervious areas such as buildings, roads and parking lots away from soil with high infiltration rate<br>• Disconnecting so that impervious areas are not continued<br>• Distributing stormwater runoff to hydrological soil type A and B (moderate to high permeability) |
| Manage stormwater runoff sources | • Applying Integrated Management Practices (IMPs)<br>• Integrating stormwater management functions into property design to create an attractive landscape that preserves the environment<br>• Creating a landscape that delays stormwater runoff and increases rainwater retained inside the property<br>• Designing so that retention and infiltration are possible in the phase of stormwater runoff source |

(*Continued*)

**Table 10.2** Strategies for LID approach (*Continued*).

| Strategy | Content |
|---|---|
| | • Integrated stormwater management combining small wetlands, permeable pavements and green roofs<br>• Distributing pollution reduction facilities such as rain garden, infiltration trench, roof cistern and ponds in runoff points<br>• Improving reliability of stormwater management system and reducing possibility of failure by comprehensively using various facilities<br>• Reducing dependence on stormwater drainage pipes, sewer pipes and ponds used in conventional stormwater runoff management<br>• Avoiding installation of stormwater drainage pipe, curbs, and runoff bypass pipe between roads and sidewalks<br>• Preventing pollution by installing best management practices (BMPs) in runoff sources |
| Maintenance and training | • Important for all stakeholders to understand and be educated<br>• Important to understand the system in which runoff is controlled by integrating with landscape and scenery with LID and to know how to manage it<br>• Need to foster housing or building owners and designers for appropriate maintenance of LID facilities<br>• Property owners must believe that these landscape elements enhance their property value, feel rewarded for contributing to environmental protection, and be willing to pay the costs required for landscape maintenance<br>• Need to develop long-term maintenance plans with clear and practicable guidelines |

*Source:* National Institute of Environmental Research [NIER] (2012).

## 10.2.2 Integrated management practices (IMPs) for LID

Stormwater IMPs include controls, operation and maintenance procedures that can be applied before, during and after precipitation events and snowmelt to reduce or eliminate the introduction of pollutants into receiving waters (US EPA, 2000). IMPs are engineered and constructed systems that are used to treat stormwater at either the point of generation or the point of discharge to either the storm sewer system or to receiving waters (US EPA, 2002, 2005). Typically, IMPs can achieve stormwater management goals by using the basic elements of flow control, detention/retention, filtration/infiltration, and biofiltration (Table 10.3). These elements can be implemented either alone or in combination, depending on site and other conditions (Wanielista & Yousef, 1993). Several categories of IMPs can be quite effective at reducing the overall volume of runoff. Volume reduction IMPs have a filtering, infiltration, biological uptake, or storage and reuse component that permanently removes some volume of runoff from the outflow (Table 10.4). IMPs that reduce volume are also reducing pollutant loads, although a concentration-in versus concentration-out study would not account for this. For this reason, the

**Table 10.3** Classification and types of IMPs for LID technological approach.

| Classification | Description | As Primary Function | As Ancillary Function |
|---|---|---|---|
| Flow control | Intercepts and regulates stormwater runoff flow rates | • Hydrodynamic device<br>• Vegetated swale<br>• Vegetated filter strips<br>• Rock swale | • Bioretention/rain garden<br>• Constructed wetland<br>• Permeable pavement<br>• Vegetated roofs |
| Detention | Provides temporary storage of stormwater runoff for the reduction of runoff volume and flow rates | • Detention pond<br>• Underground detention tank<br>• Dry swale | • Bioretention/rain garden<br>• Constructed wetland<br>• Permeable pavement<br>• Vegetated swale<br>• Vegetated filter strips<br>• Rock swale |
| Retention | Provides storage of stormwater runoff to allow settling for the removal of particulate materials | • Retention basin<br>• Rainwater harvesting | • Bioretention/rain garden<br>• Constructed wetland |
| Filtration | Filters stormwater runoff by using porous media (e.g., sand, natural or synthetic filter materials, geotextile fabric, etc.) for partial treatment of stormwater runoff | • Tree box filter/planter<br>• Surface sand filter<br>• Underground sand filter | • Bioretention/rain garden<br>• Constructed wetland<br>• Infiltration trench<br>• Infiltration basin<br>• Permeable pavement<br>• Vegetated swale<br>• Vegetated filter strips<br>• Rock swale |
| Infiltration | Infiltrates stormwater runoff underground enhancing groundwater recharge | • Infiltration trench<br>• Infiltration basin<br>• Permeable pavement | • Bioretention/rain garden<br>• Constructed wetland<br>• Vegetated roof |
| Water quality control | Utilizes chemical and biological processes by means of plants, soil and microorganisms for the treatment of pollutants | • Bioretention/rain garden<br>• Constructed wetland (free water surface flow, horizontal/vertical subsurface flow, hybrid) | • Tree box filter/planter<br>• Vegetated swale<br>• Vegetated filter strips<br>• Infiltration trench<br>• Infiltration basin<br>• Vegetated roof |

removal efficiency of these types of IMPs may be under-reported, especially when a concentration-in versus concentration-out study approach was used.

**Table 10.4** Treatment removal processes occurring in IMPs.

| Pollutant | Removal Process in LID Practice | | | | |
|---|---|---|---|---|---|
| | **Ponds** | **Wetlands** | **Infiltration** | **Biofilter** | **Sand Filter** |
| Solids | Settling | Adsorption Settling | Adsorption Filtration | Adsorption Filtration Settling | Filtration |
| Nutrients | Biological uptake | Biological uptake | Adsorption | Adsorption Biological uptake | Adsorption |
| Heavy metals | Adsorption Settling | Adsorption Settling Biological uptake | Adsorption Filtration | Adsorption Filtration Settling Biological uptake | Adsorption Filtration |
| Toxic organics | Adsorption Degradation Biological conversion | Adsorption Degradation Biological conversion | Adsorption | Adsorption Filtration Biological conversion | Adsorption Filtration |
| Oil and grease | Adsorption Settling | Adsorption Settling | Adsorption | Adsorption Settling | Adsorption |
| Biological Oxygen Demand | Biological conversion | Biological conversion | Biological conversion | Biological conversion | Biological conversion |
| Bacteriological | Settling UV (sunlight) Predation | Settling UV (sunlight) Predation | Filtration | Filtration Settling | Filtration |

*Source:* Adapted from Scholze *et al.* (1993).

Effective low impact development is not simply a matter of selecting from a menu of available preferred practices. Rather, it is an integrated planning and design process. It includes the use of stormwater management measures that focus first on minimizing both the quantitative and qualitative changes to a site's pre-developed hydrology and then providing treatment as necessary through a network of facilities distributed throughout the site. Though one facility alone will likely not satisfy performance requirements, facilities with varying levels of service in a treatment network will provide superior levels of treatment and volume reduction (University of Arkansas [UOA], 2010).

## 10.2.3 Sizing criteria of LID facility

The hydrologic approach of LID is to retain the amount of rainfall within the development site as was retained prior to any development; that is to mimic the pre-development hydrologic regime to maximize protection of receiving waters, aquatic ecosystems and groundwater recharge. The required volume of LID practices can be determined by several methods. However, it is important to note that the surface storage must contain the required water quality captured volume (fixed volume above the soil media surface). The uncaptured runoff volume should be bypassed to another device/facility if treatment of the entire runoff from the site is desired. Ideally, LIDs should be sized and designed to provide both water quality control and peak flow control.

#### 10.2.3.1 Water quality volume (WQV)

The most widely used sizing criterion for LID practices is the water quality volume (WQV) estimated as the product of rainfall, runoff coefficient and catchment area (Eqn. 1). WQV pertains to the amount of stormwater runoff from a rainfall event that should be captured and treated to remove the majority of stormwater pollutants on an average annual basis. Recent research suggests that the most economical and effective water quality sizing criterion is the first flush runoff depth. Other water quality sizing methods were developed to achieve higher pollutant removals, in which the WQV is equal to the storage required to capture and treat 90% of annual runoff and pollutant load (Kang *et al.* 2006; Lee *at al.*, 2010; Shafiri *et al.* 2011). Many LID sites are designed to treat runoff from small storms, rather than large storms, so it is important to understand the basis of design when applying on a catchment-scale basis. The WQV sizing method is generally used for sizing the LID practices of retention and infiltration types such as constructed wetland, retention pond, bioretention, infiltration trench, etc.

The WQV is estimated based on the accumulated effective rainfall using the equation below:

$$\text{WQV} = P \times C \times A \tag{10.1}$$

Where, WQV = water quality volume (m$^3$), $P$ = accumulated effective rainfall (mm), $C$ = runoff coefficient (unitless), $A$ = total catchment area draining to the LID (m$^2$), Figure 10.2.

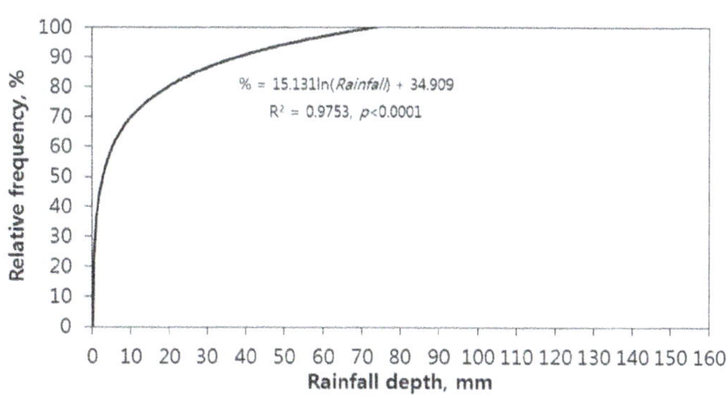


**Figure 10.2** Sample rainfall frequency curve for WQV.

### 10.2.3.2 Water quality flow (WQF)

Flow-based treatment LID practices are designed to treat higher flow rates, and do not necessarily require a huge storage volume. Thus, these practices are more appropriately designed based on peak flow rate, rather than water quality volume. The water quality flow (WQF) is used to estimate the flow rate associated with the WQV, for sizing flow-based treatment and pre-treatment practices (Eqn. 2).

$$\mathrm{WQF} = C \times I \times A \tag{10.2}$$

Where, WQF = water quality flow ($m^3$/hr), $C$ = runoff coefficient (unitless), $I$ = design rainfall intensity (mm/hr), $A$ = total catchment area draining to the LID ($m^2$), Figure 10.3.

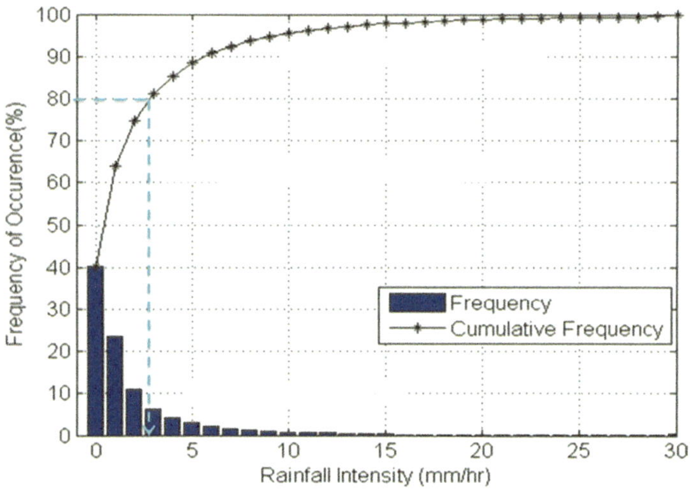


**Figure 10.3** Sample rainfall intensity frequency curve for WQF.

### 10.2.3.3 Groundwater recharge volume (GRV)

Another volume-based sizing criterion is the inclusion of groundwater recharge volume (GRV) to the WQV. The GRV is considered as part of the WQV to protect groundwater resources by maintaining the pre-development groundwater discharge volumes. Maintaining pre-development groundwater recharge conditions can also reduce the volume of runoff that must be managed to meet other design criteria (i.e., water quality, channel protection, and peak flow control), and thus the overall size and cost of stormwater management practices. The GRV from post-development impervious surfaces can be achieved by means of capturing and infiltrating a portion of runoff from larger storms, or runoff from the most frequent small storms using infiltration/filtration LID practices. The GRV is a function of annual pre-development recharge rate for site specific soil condition, annual rainfall and impervious cover on a site (Eqn. 3).

$$\mathrm{GRV} = A \times GR \tag{10.3}$$

Where, GRV = groundwater recharge volume ($m^3$), $A$ = total effective area of impervious surfaces that will exist on the site after development ($m^2$), $GR$ = groundwater recharge depth based on the hydrologic soil group (m).

The hydrologic soil group approach uses estimates of average annual infiltration rates for each hydrologic soil group. For each hydrologic soil group, the groundwater recharge depth is the amount of runoff that must be captured from an impervious surface and infiltrated for each storm, in order to make up for the loss of recharge that would otherwise result from that impervious surface. The cumulative effect of capturing and infiltrating the initial volume of runoff from multiple events is to approximate the annual recharge occurring during pre-development conditions.

## 10.2.4 Pretreatment and sediment control

Particles washed-off from the paved areas contain various sorbed pollutants, and much of the pollutant load associated with stormwater runoff is carried by sediment. Thus, existing technologies for treating stormwater runoff are often targeted at removing pollutants bound to particles. In fact, many LID practices rely on sedimentation as the primary pollutant removal mechanism therefore pretreatment is necessary. Pretreatment is also required to both minimize groundwater contamination and to prolong the life of the LID facilities. Figure 10.4 shows the primary pretreatment functions and examples and pretreatment practices.

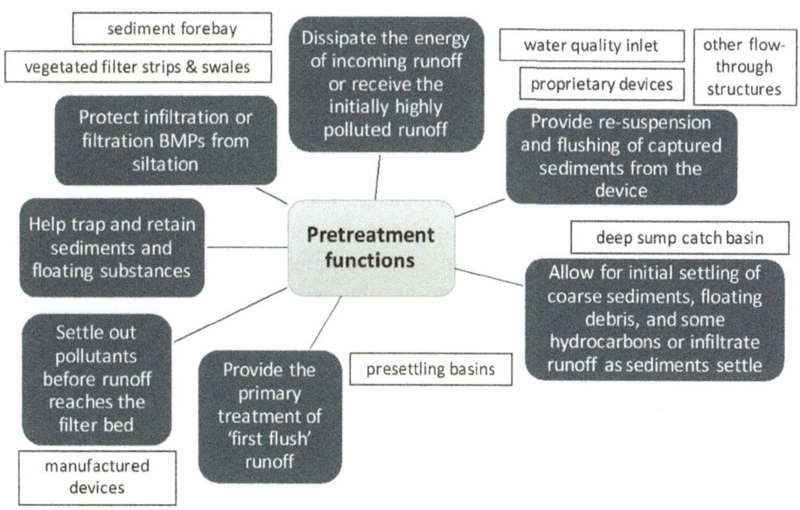


**Figure 10.4** Primary functions and examples of pretreatment practices.

Presettling basins that are often integrated into the design of stormwater management structures provide the primary treatment of 'first flush' runoff. In Korea, use of pre-settling basins is a popular pretreatment practice, especially applied when the proposed treatment site is adversely affected by a pollutant load with a hydrology condition of dominantly low-rainfall frequency occurrence storm events (Maniquiz-Redillas, 2014).

#### *10.2.4.1 Pretreatment considerations:*

- Settling will take place after stormwater is trapped and ponded between storms.
- The efficiency of a presettling basin in an LID practice in removing particulates by settling is dependent upon the initial concentration of suspended solids in the runoff.
- Some particles, such as fine clays, will not settle out of suspension without the aid of a coagulant. Turbulence, eddies, circulation currents, and diffusion at inlets affect the settling ability of particles.
- Pollutants such as metals, hydrocarbons, nutrients, and oxygen-demanding substances can become adsorbed or attached to particulate matter, particularly clay soils. Removal of these particulates by sedimentation can therefore result in the removal of a large portion of these associated pollutants.
- Sediments and associated pollutants captured in the pretreatment devices can be removed for disposal during maintenance operations.

## 10.3 HYDROLOGICAL AND ENVIRONMENTAL EFFECTS

### 10.3.1 Changes of water quality and runoff flow after LID application

The LID approach combines a hydrologically functional site design with pollution prevention measures to compensate for land development impacts on hydrology and water quality. LID application in urban areas plays the role of recovering natural water circulation and reducing pollutants discharged from pavements. Recovery of natural water circulation can be achieved by increasing the infiltration ability of soil and evapotranspiration of plants. Recovering natural water circulation through LID installation is possible by delaying the runoff time after or during rainfall, reducing runoff volume and peak flow, and delaying peak flow occurrence time. Reduction of stormwater runoff volume through LID application may reduce the outflow of pollutants. Pollutants are reduced through physical reduction mechanisms such as biochemical reduction mechanisms, filtration and absorption of microorganisms and plants inside the soil. Figure 10.5 exhibits the stormwater runoff flow rates and pollutant concentrations before and after the application of an infiltration trench on a highly impervious paved area (Maniquiz, 2012).

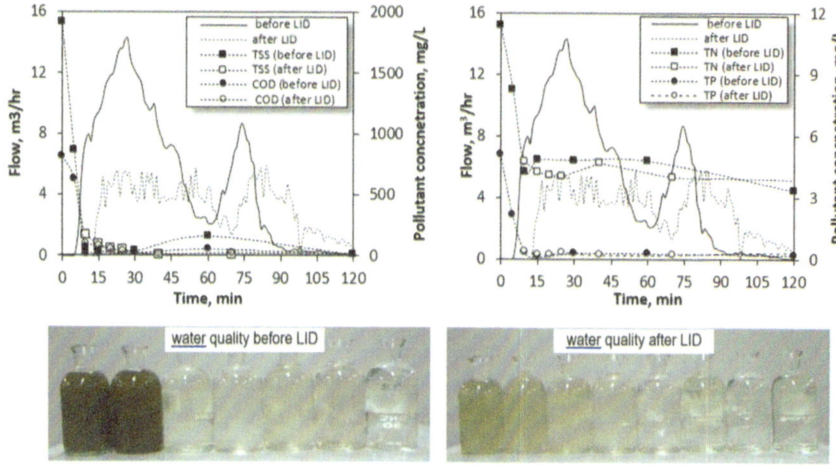


**Figure 10.5** Water quality before and after the application of a one-year old infiltration trench for August 10, 2010 storm event (total rainfall depth = 22.5 mm, rainfall duration 3.32 hr, antecedent dry period = 2.4 days).

## 10.3.2 Effect on the volume reduction after LID application

As the imperviousness of the site is increased, the runoff volume for a given storm increases. The ratio of the corresponding amount of runoff to the total amount of rainfall or precipitation is called the runoff coefficient. The typical site runoff coefficient can be maintained at the pre-development level by compensating for the loss of abstraction (interception, infiltration, depression storage) through both site planning and design considerations.

Peak flow reduction and delay of occurrence time have the effect of improving water quality and aqua-ecosystem by supplying ecological water to the streams, while also reducing flooding of urban areas. Therefore, advanced countries including the United States and Korea are currently considering the use of LID/GI technologies as disaster prevention facilities to reduce urban flooding. As shown in Figure 10.6, most LID/GI facilities show the reduction in peak flow, as well as delaying the peak flow occurrence time.

## 10.3.3 Effect on the pollutant reduction after LID application

LID is typically designed to provide water quality treatment control for desired runoff depth or volume from impervious areas using retention practices. In most LID applications, the use of distributed control and retention throughout the site will result in much higher levels of water quality treatment control for a number of reasons. First, the runoff volume controlled will treat a greater volume of annual runoff usually associated with decreases in both the time of concentration and flow velocities, which results in a reduction in the pollutant transport capacity and overall pollutant loading. In Figure 10.7, the influence of rainfall range depth in the reduction of pollutant load by an infiltration trench designed using 25 mm rainfall (WQV), is emphasized. For a rainfall range less than 20 mm, the pollutant reduction was all positive and high between 55 and 75% except for total lead (Tot-Pb). However, negative reduction was observed for a rainfall range greater than 20 mm for most heavy metals. LID also supports pollution prevention practices by modifying human activities to reduce the introduction of pollutants into the environment.

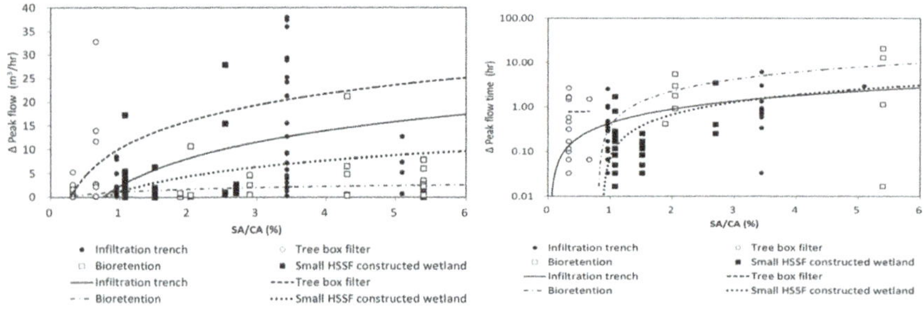


**Figure 10.6** Peak flow reduction with surface area/catchment area (SA/CA) after LID/GI application.

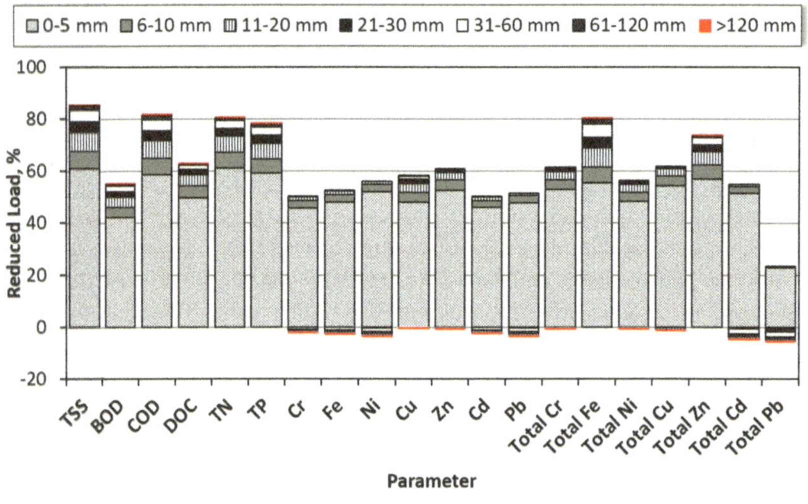


**Figure 10.7** Estimated proportion of accumulated reduced runoff pollutant load for each respective rainfall depth range at an infiltration trench (Data from over 27 storm events from May 2009 to February 2013).

Figure 10.8 shows the load reduction for BOD and total nitrogen (TN) from roof runoff after the application of a rain garden estimated with respect to rainfall depth using different surface area to catchment area (SA/CA) ratios (Maniquiz, 2012). As can be seen, higher pollutant load reduction could bc achieved for smaller rainfall depth. There was no specific trend in which the pollutant load was greatly reduced based on SA/CA ratio. However, it was observed that the rain garden base design of 5% SA/CA was nearer the performance to SA/CA of 7% than 3%. Nevertheless, selection of the SA/CA ratio depends on the requirement of load reduction that should be specified in the regulations.

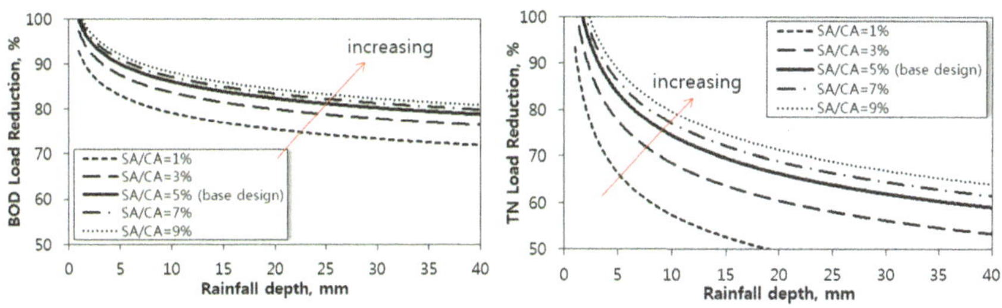


**Figure 10.8** Pollutant load reduction with respect to rainfall depth for various SA/CA ratios calculated on rain garden performance data collected from April 2011 to February 2013.

### 10.3.4 Ancillary effects and benefits of LID application

LID has numerous benefits and advantages and is a more environmentally sound technology. LID can enhance the local environment and protect public health, environmental assets and water quality; and builds community livability. Natural functions can be maintained with the use of LID practices, which include reduced impervious surfaces, functional grading, open channel sections, disconnection of hydrologic flow paths, and the use of bioretention/filtration landscape areas (Coffman, 2002; CSD, 2007). LID practices increase natural rainfall penetration and natural groundwater recharge, thus reducing potential impacts on biological habitat and reduced base flow into reservoirs from extended drought periods (CSD, 2007; Gilroy & McCuen, 2009). The natural processes employed by LID practices allow pollutants to be filtered or biologically or chemically degraded before stormwater reaches the water bodies (CSD, 2007).

Figure 10.9 shows the difference in surface temperature between an asphalt and permeable pavement. The use of permeable pavement contributed to a decrease in surface temperature in urban areas that could possibly lead to a partial reduction of the heat island effect. Retention and infiltration of urban water using LID facilities also produce a variety of effects such as reduction of pollutants, recovery of water circulation, provision of ecological habitats, prevention of urban disasters, and reduction of energy use.

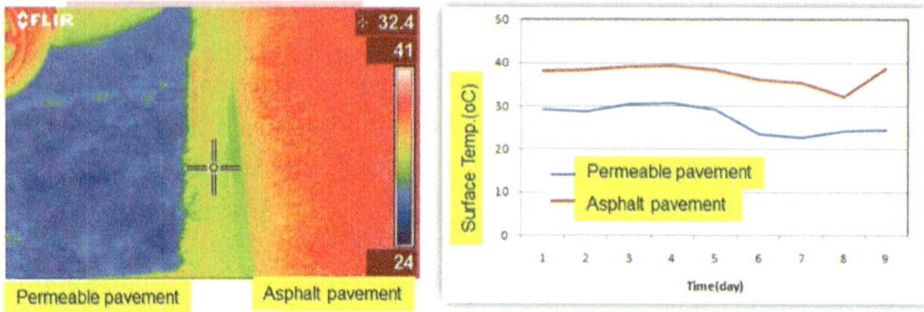


**Figure 10.9** Surface temperature reduction by permeable pavement.

The opportunities, effectiveness, and benefits for control of runoff through numerous small-scale multifunctional landscape features have not been fully explored. To apply LID in industrial sites is simply a matter of developing numerous ways to creatively prevent, retain, detain, use, and treat runoff within multifunctional landscape features unique to that site.

## 10.4 CONCLUSION

Toxic substances such as acids, alkalis, degradable organic residues, detergents, disinfectants, dyes, engine coolants, fertilizers, fuel, lubricants, metal solutions, pharmaceuticals, salts, poisons and solvents present in industrial sites could be transported by stormwater runoff and can cause a minor inconvenience to a major disaster bringing harm to people, property and/or ecosystems. In order to avoid flooding and contamination risks in industrial areas, LID practices are widely applied in developed countries to minimize the hydrological and environmental impacts of stormwater runoff. The basic LID strategy to manage stormwater is to reduce runoff volume and decentralizing flows, best accomplished by creating a series of smaller retention/detention areas that allow localized filtration instead of carrying runoff to a remote collection area to be treated. As nonpoint source pollution continues to be the focus of watershed management within municipalities, development and implementation of effective stormwater management practices has emerged as the key to controlling this inherently diffused and decentralized source. In particular, implementation of stormwater IMPs into LID in industrial areas undergoing development and redevelopment is recommended to focus on minimizing post-development peak discharge rates, volume of runoff and pollutant loads, to mimic pre-development values with the ultimate goal of protecting and/or improving the quality of receiving waters.

## REFERENCES

Coffman L. S. (2002). Low Impact Development: Smart Technology for Clean Water Definitions, Issues, Roadblocks, and Next Steps. Urban Drainage 2002: 9th International Conference on Urban Drainage. Sept. 8–13, 2002. Portland, Oregon.

County of San Diego (CSD) (2007). Low Impact Development Handbook – Stormwater Management Strategies. Department of Planning and Land Use, San Diego, California. December 2007.

Davis A. P. (2005). Green engineering principles promote low-impact development. *Environmental Science & Technology*, **39**(16), 338–344.

Ferguson B. (2002). Stormwater management and stormwater restoration. In: Handbook of Water Sensitive Planning and Design, R. L. France (ed.), Lewis Publishers, Washington DC.

Fletcher T. D., Shuster W., Hunt W. F., Ashley R., Butler D., Arthur S., Trowsdale S., Barraud S., Semadeni-Davies A., Bertrand-Krajewski J.-L., Mikkelsen P. S., Rivard G., Uhl M., Dagenais D. (2015). SUDS, LID, BMPs, WSUD and more – The evolution and application of terminology surrounding urban drainage. *Urban Water Journal*, **12**(7), 525–542.

Gilroy K. L. and McCuen R. H. (2009). Spatio-temporal effects of low impact development practices. *Journal of Hydrology (Amsterdam)*, **367**(3–4), 228–236.

Guillette A. (2007). Achieving sustainable site design through low impact development practices. In: Whole Building Design Guide. National Institute of Building Sciences, Washington, DC.

Kang J. H., Kayhanian M. and Stenstrom M. K. (2006). Implications of a kinematic wave model for first flush treatment design. *Water Research*, **40**(20), 3820–3830.

Lee E. J., Maniquiz M. C., Gorme J. B. and Kim L. H. (2010). Determination of cost-effective first flush criteria for BMP sizing. *Desalination and Water Treatment*, **19**, 157–163.

Lloyd S. D., Wong T. H. F. and Chesterfield C. (2002). Water Sensitive Urban Design: A Stormwater Management Perspective. CRC for Catchment Hydrology Industry Report 02/10.

Maniquiz M. C. (2012). Low Impact Development (LID) Technology for Urban Stormwater Runoff Treatment–Monitoring, Performance and Design. PhD thesis, Kongju National University, Department of Civil and Environmental Engineering, Cheonan City, South Korea.

Maniquiz-Redillas M. C., Geronimo F. K. F. and Kim L. H. (2014). Investigation on the effectiveness of pretreatment in stormwater management technologies. *Journal of Environmental Sciences*, **26**(9), 1824–1830.

Mull K. K. (2005). Selling low impact development: audiences, messages, and media. In: Fourth National Nonpoint Source and Stormwater Pollution Education Programs. Holiday Inn, Chicago, pp. 46–52.

National Institute of Environmental Research (NIER) (2012). The Technical Guideline of Water Pollutant Load Management. National Institute of Environmental Research, Incheon South Korea [Korean Literature].

Prince George's County (PGC) (1999). Low-Impact Development Design Strategies: An Integrated Design Approach. Prince George's County, MD Department of Environmental Resources. June 1999.

Puget Sound Action Team (PSAT) (2005). Low Impact Development: Technical Guidance Manual for Puget Sound. Publication No. PSAT 05–03. Washington State University Pierce County Extension, Olympia, WA.

Scholze R., Novotny V. and Schonter R. (1993). Efficiency of best management practices for controlling priority pollutants in runoff. *Water Science and Technology*, **28**(3–5), 215–224.

Sharifi S., Massoudieh A. and Kayhanian M. (2011). A stochastic stormwater quality volume-sizing method with first flush emphasis. *Water Environment Research*, **83**(11), 2025–2035.

U.S. Environmental Protection Agency Office of Water (US EPA) (2005). National Management Measures to Control Nonpoint Source Pollution from Urban Areas. EPA- 841-B-05–004. Office of Water, Washington DC.

U.S Environmental Protection Agency (US EPA) (2000). NPDES National Menu of Stormwater Best Management Practices. http://cfpub.epa.gov/npdes/stormwater/menuofbmps/

U.S. Department of Defense (US DoD) (2004). Unified Facilities Criteria – Design: Low Impact Development Manual, Unified Facilities Criteria No. 3–210–10. U.S. Army Corps of Engineers, Naval Facilities Engineering Command, and Air Force Civil Engineering Support Agency.

U.S. Environmental Protection Agency (US EPA) (2002). Considerations in the design of treatment best management practices (BMPs) to improve water quality. USEPA report EPA/600/R-03/103.

University of Arkansas (UOA) (2010). Low Impact Development: A Design Manual for Urban Areas. Fay Jones School of Architecture, University of Arkansas Press.

Wanielista M. P. and Yousef Y. A. (1993). Stormwater Management. J. Wiley & Sons, New York, p. 579.

# Chapter 11

# The application of sustainable drainage technology: challenges and solutions

*C. Pittner[1]*, N. S. Campbell[2], H. Sommer[3], H. Sieker[3], B. J. D'Arcy[2]*

*[1]WSP, Parsons Brinckerhoff, 7 Lochside View, Edinburgh Park, Edinburgh*

*[2]C&D Associates, and independent consultant, www.enviroexperience.co.uk, Edinburgh, UK*

*[3]Ingenieurgesellschaft, Hoppegarten, Near Berlin*

**Corresponding author: chris.pittnerspgroup.com*

## 11.1 INTRODUCTION

Infrastructure requirements for pollution prevention and flood risk management have been set out in text books and guidance for many years, for example Horner *et al.* (1994), Urbonas and Stahre (1993), Schueler (1987), Schueler *et al.* (1992), CIRIA (2000, 2015). The techniques detailed are variously described in the literature and in policy guidance in each country as urban best management practices (BMPs), sustainable urban drainage systems (SUDS) and elsewhere as Low Impact Development (LID) techniques. Such guidance however, is not enough on its own to deliver fit-for-purpose application of the technology. Case studies of successful application of the technical guidance are valuable, since different constraints can be described and how they were overcome demonstrates the ways in which guidance can be effectively translated into reality. In addition, although all three examples here are excellent, well respected demonstration developments,

© IWA Publishing 2017. Wealth Creation without Pollution: Designing for Industry, Ecobusiness Parks and Industrial Estates
Brian D'Arcy, Lee-Hyung Kim, Marla Maniquiz-Redillas
doi: 10.2166/9781780408330_153

small examples of details that could be better done elsewhere in subsequent developments can be used to indicate risks when introducing new ideas, and avoid similar problems in future. Here three European case studies demonstrate different challenges according to local regulatory context, soil type, pollution and flood risks, the drainage techniques chosen, and the scale of development. For convenience, the stormwater management technology for minimising adverse environmental impacts is referred to here for each case study, as sustainable drainage systems (Stahre, 2006), or SUDS (Campbell *et al.* 2004; D'Arcy, 2013).

## 11.2 THREE EUROPEAN CASE STUDIES

For each case study example, critical site details (soils, area in hectares, percentage imperviousness, infrastructure design and provision, land-use, with consequent pollution activities present and risks, are indicated.

The three case studies are:

- J4M8 Distribution Park, Near Bathgate, West Lothian, Scotland UK
- A major distribution depot within the Dunfermline East Expansion (DEX) site; (Amazon depot), Scotland UK
- Hoppegarten Industrial estate, near Berlin, Germany.

The first two sites are in the Central Belt of Scotland, UK, and the third site is just east of Berlin, Germany. A summary of key features of each case study is given in Table 11.1. The types of SUDS features used in each instance, and the pollution risks they address, are indicated in summary in Table 11.2. All three are examples of new build developments, and none are in free draining porous soils that would be simple for infiltration.

## 11.3 ENGINEERING CONSIDERATIONS

Engineering considerations – objectives and challenges and how value is added to the basic requirement to drain the premises. Design storm, flow controls, response to the soil types, return period, and winter water table, are compared for the three case studies.

### 11.3.1 J4M8 Distribution park, near Bathgate, West Lothian

Drainage and SUDS design for the 83-hectare distribution park began in 2001 and the site was completed in 2008 (Figure 11.1). SEPA (Scottish Environment Protection Agency) policy specified that the SUDS treatment train required three levels of treatment for this type of development: source control within plot boundaries; site control using swales/linear wetlands; and estate/regional control with retention ponds.

**Table 11.1** Comparison of the three case study example developments.

| Design Considerations | J4M8, near Bathgate, UK | DEX, Dunfermline, UK | Dahlwitz-Hoppegarten, near Berlin, Germany |
|---|---|---|---|
| Type of development | Distribution park, close to the M8 motorway linking Edinburgh and Glasgow | Distribution depot, within a mixed industrial, commercial, housing & retail | Industrial/ commercial estate (green field development, plus some existing) |
| Development area | 83 hectares | DEX is 5 $km^2$ of which the distribution depot is 20 ha | 160 ha of which 40 ha is already existing |
| Soil type | Glacial till | Relatively deep topsoil from previous arable farmland use over glacial till/boulder clay | Glacial loamy soil |
| Soil porosity | $10^{-9}$ m/s or lower | Effectively highly impermeable | Infiltration rate $5*10^{-6}$ m/s |
| Local watercourse | Culverted tributary of R Almond | Lyne Burn – history of flooding | Wernergraben stream |
| Sewer system | New separate sewers | New separate drainage but old town on combined sewers, with history of surcharging | Conventional sewers on existing 40 ha (a quarter of the development) |
| Drainage capacity | Phase 1 drainage required to pass through existing small drain that passes beneath M8 motorway | Limited due to history of flooding in Dunfermline prior to development of agricultural headwaters | 40 l/s permitted, approx. natural runoff rate |
| Groundwater | >2 m below ground level | Naturally wet landscape | In wet years, temporary groundwater at surface |
| Topography | Relatively flat | Rolling glacial deposits; moraine hills | Relatively flat |
| Previous land-use | Livestock grazing | Mixed arable and rough grazing farmland | Undeveloped |

**Table 11.2** Pollution risks and types of SUDS provided at the three case study example developments.

| Pollution Risks and SUDS Types | J4M8, near Bathgate UK | DEX, Dunfermline, UK | Dahlwitz-Hoppegarten, near Berlin, Germany |
|---|---|---|---|
| Pollution risks | Spillage, leakage and vehicle washing & rainfall mobilised atmospheric deposits | Vehicle washing & rainfall mobilised atmospheric deposits | Vehicle washing & rainfall mobilised atmospheric deposits, variety of site specific pollutants |
| Pollutants | Oil and PAHs & detergents, beer, toxic metals | Oil and PAHs & detergents, toxic metals | Oil and PAHs & detergents, toxic metals |
| Filter drains | Limited: as source controls in some of the premises | On approach roads and within premises | – |
| Under-drained swales | – | – | Widespread as roadside features |
| Reedbed | Only as margins of stormwater wetland | Only as margins of stormwater wetland | Site control at one of the premises |
| Grass swale | Main means of conveyance from individual premises to regional feature | Only as feed from gravel drainage network and site runoff outflow into pond | Provided on several of the individual premises |
| Permeable pavement | As source control at some premises | Open pore blacktop on car park | Various on some individual premises |
| Detention pond or basin | 2 ponds as regional features for phase 1 and 2 respectively | One pond serving whole site plus approach roads | Extended detention area serves whole estate |

### *11.3.1.1 Objectives and challenges*

The main objectives and challenges were:

Meeting West Lothian Council's policies and requirements with respect to attenuation and designing for exceedance

Designing SUDS to be integrated with their surroundings

Design taking cognisance of location of underground and overhead services

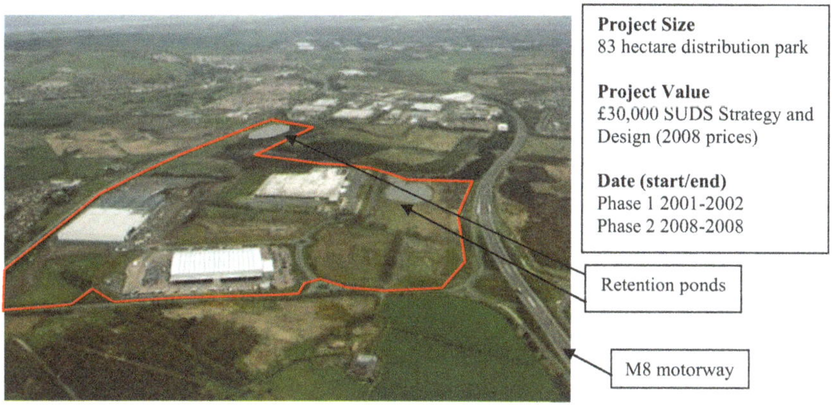


**Figure 11.1** J4M8 Distribution park, West Lothian, Scotland, UK. *Photo credit: Chris Pittner 07/2008.*

### 11.3.1.2 Added value and benefits

| | |
|---|---|
| Practical design bringing 20% cost saving on original budget estimate | Visible System which is easily maintained |
| Effective working relationship with SEPA and West Lothian Council | Use of swales reduces extent of piped system saving on construction costs. |
| Maximising developable land whilst maintaining SUDS requirements | Added amenity and ecological value |

### 11.3.1.3 Solutions

The design solution to meet the requirements of the SUDS treatment train included design of approximately 700 m of swale/linear wetland, providing the second level of treatment and a retention pond providing 4 × Water Treatment Volume (*Vt*) and attenuation for up to the 200-year return period storm event, providing the third level of treatment. A penstock valve on the outlet from the retention ponds, allows for closure, and capture of floating material too. Source control first level of treatment is provided by the occupiers of each plot within the estate.

The phase 2 drainage features were designed on the following basis:

The areas within each plot boundary, abstracted from the site master plan, were assumed to be impermeable, yielding a 'worst case' scenario for attenuation calculations. The overall impermeable area of the contributing plots, was calculated to be approximately 21.4 ha. The treatment volume, *Vt*, calculated for this area is 82 $m^3$/ha, giving a total for the site area of 3034 $m^3$. SEPA and West Lothian Council intimated that a volume of 4 × *Vt* (12,136 $m^3$) would be required for a retention pond and 3 × *Vt* if a wetland treatment system was implemented.

The retention pond was sized so that the design treatment volume can be fully contained, with the outlet being designed so that the attenuation storage volume emptied over a period of no less than 24 hours.

The treatment volume, *Vt*, as described in CIRIA C697 The SUDS Manual, is based on an empirical formula linked to the M5–60 rainfall depth for the area, equivalent to approximately 10–20 mm rainfall depths, and has been calculated using the following parameters:

*D* = mm rainfall depths of five-year return period storm of 60 minutes duration

SOIL = soil index broadly describes the infiltration potential

*i* = assumed impervious fraction for development

$$Vt = 9D\left[\frac{\text{SOIL}}{2} + \left(1 - \frac{\text{SOIL}}{2}\right)i\right] \text{m}^3/\text{hectare of development site area}$$

Attenuation is a West Lothian Council requirement in accordance with their requirement to control flooding.

As the site discharged to a watercourse, via a culvert, the council required the surface water discharge from the post-development site to be attenuated to, at most, the 2-year return period storm green-field runoff rate from the existing site area.

The 2-year return period rainfall event pre-development runoff was calculated in accordance with guidance given in CIRIA C697 The SUDS Manual – 'estimating greenfield runoff rates' using the following formula from the Institute of Hydrology Report 124:

$$\text{QBARrural} = 0.00108\text{AREA}^{0.89}\ \text{SAAR}^{1.17}\ \text{SOIL}^{2.17}$$

Where:

QBARrural = Catchment mean annual peak flow ($m^3/s$)
AREA = Catchment area ($km^2$)
SAAR = Standard average annual rainfall for the period 1941–1970 (mm)
SOIL = Soil Index (from Flood Studies or Wallingford Procedure WRAP maps)

West Lothian Council indicated that full attenuation would be required up to the 30-year return period storm, with a provision to accommodate the 200-year return period storm within the development site area without detriment to properties within or out-with the proposed site area.

The results of the retention pond modelling indicated that the volume required to provide the necessary attenuation for the 200-year period storm was 16,500 $m^3$. This related to a winter storm of 72 hours duration. The maximum outflow for this storm was calculated to be 68.8 l/s, which was compliant with the limiting discharge requirements.

## 11.3.2 Dunfermline east expansion site, Dunfermline, Scotland UK

The Amazon distribution centre in Dunfermline was developed by BWB Consulting Ltd., who undertook drainage and SUDS design for the 20-hectare distribution centre. It lies within the Duloch Park district of the 5 $km^2$ Dunfermline East Expansion site (DEX). Work started on site in 2011 and construction was completed in 2012. The area drains to the Lyne Burn, a small stream, which flows through the old town of Dunfermline downstream of the DEX site. In addition to the need to prevent pollution from new development, a history of flooding in the old town meant that flood risk minimisation was a priority for the drainage infrastructure provided for the DEX development from the outset. Accordingly, a drainage master plan was produced in 1995/1996 (Roesner & Campbell, 1996). Guidance developed there led directly to SEPA policy (SEPA, 1996), and the first UK SUDS manual (CIRIA, 2000). For industrial estates that specified that the SUDS treatment train required three levels of treatment, these were: at source, conveyance, and a final 'polishing' and flow attenuation retention pond or constructed wetland.

### *11.3.2.1 Objectives and challenges*

- Meeting the policies and required development standards of Fife Council, with respect to attenuation and designing for exceedance, as outlined in the Fife Council Flood Liaison and Advise Group Guidance Note on Flooding and Drainage Issues in Relation to Planning and Development (2006)
- Compliance with the drainage master plan for the development (Roesner & Campbell, 1996)
- Designing SUDS to be integrated with their surroundings
- Design taking cognisance of location of underground and overhead services, and receiving water and drainage infrastructure and capacity (the new development had to fit within a 15-year-old DEX development and strategic drainage system)

### *11.3.2.2 Added value and benefits*

| | |
|---|---|
| Practical design featuring variety of source control techniques | Fenced pond provides safety for local children, but also excludes them and dogs, thereby allowing birds and other wildlife to breed there free from disturbance |
| Effective working relationship with SEPA and Fife Council | Landscaping plan, with higher level concrete access road within the depot, allowed for runoff from the impermeable road to drain down into the permeable surfaced car park below, minimizing drainage requirements |
| Final pond for balancing high flows and for 3rd stage treatment meets SUDS requirements | Efficient use of space achieved by integration with landscaping requirements, and use of traffic island on approach road as a detention basin |

#### *11.3.2.3 Solutions*

The design solution to meet the requirements of the SUDS treatment train included design of approximately 700 m of filter drain/underdrained swale, a detention basin within a traffic island, and a swale/linear wetland, providing the second level of treatment draining to a retention pond, providing 4 times the Water Treatment Volume (*Vt*) and attenuation for up to the 1-in-100-year return period storm event, providing the third level of treatment. Source control first level of treatment provided by underdrained swales and filter drains alongside approach road and the concrete access roads for trucks within the depot, but permeable pavement for the staff and visitors car parking areas.

The drainage features were designed on the following basis:

The SUDS surface water drainage was generally designed to cope with 1-in-100-year return period storm events +30% allowance for climate change, which in post development discharge terms was to be limited to the equivalent 1-in-1-year return period flow level of 60.0 l/s, which complied with Fife Council's required development standards.

The drainage systems were also required to be designed to ensure that no back up flooding should occur within the full drainage systems in applied storm events up to and including the 1-in-200-year return period.

In terms of the SUDS features there are also requirements that Fife Council need to be satisfied relative to their landscaping; managed maintenance and the provision of measures to inhibit access for public safety.

### 11.3.3 Hoppegarten industrial estate, near Berlin, Germany

Although it was primarily a green field development, in order to prevent any increase in flood risk in the small receiving watercourse, the local planning authority required that the existing small drain serving the undeveloped site should be retained as the outlet pipe for the post-development site. The regional water authority had set a maximum yearly peak flow of 400 l/s from the development area draining into the small Wernergraben creek, and the existing developed area was already contributing 360 l/s (Sommer *et al.* 2008). The entire new development area therefore had to meet a maximum discharge limit of only 40 l/s, comparable with the natural discharge of the area.

Local soils and winter water table were not favourable for infiltration, so an under-drained swale network, utilising engineered soils overlying geocellular units was designed and installed throughout the new parts of the development. Individual premises have a variety of source control techniques, from swales and small detention basins to permeable pavement.

#### 11.3.3.1 Objectives and challenges

The main objectives and challenges were:

- Flood risk management requirements limited the maximum flow to approximately the pre-development condition
- The German water legislation (Wasserhaushaltsgesetz WHG) requires that the capability of the natural water system be preserved, including infiltration and evaporation (Sommer *et al.* 2008).
- A local planning regulation required private properties to have on-site stormwater management (Panning & Sieker, 1998).
- Pollution prevention was also a requirement

#### 11.3.3.2 Added value and benefits

Decentralised stormwater attenuation was less expensive than conventional drainage plus centralised storage:

| Costs | Centralised Retention | Decentralised Retention |
|---|---|---|
| Total cost | 19 Euros/m$^2$ | 17 Euros/m$^2$ |
| Public part | 16 Euros/m$^2$ | 5 Euros/m$^2$ |
| Private part | 3 Euros/m$^2$ | 12 Euros/m$^2$ |

The main benefits achieved were:

- Cost savings by comparison with a conventional drainage system
- Measured performance within design objectives for discharge from the development
- Attractive green infrastructure for a commercial/industrial development
- High visibility of any pollution incidents – more evident in the green infrastructure than in a conventional drainage network
- An oil spill occurred and was observed to be retained in the first 10–20 cm of the topsoil layer, where it was broken down in the aerobic conditions without need for excavation of the contaminated soil
- Longevity of the system; evaluation after 15 years showed no loss of functionality (Boogard *et al.* 2014).

#### 11.3.3.3 Solutions

The drainage solution for the area was to implement swale-trench-systems to reduce water flow to the adequate rate (Figure 11.2). For the conventionally drained areas, built when retention was not required or built with storage sewers, additional semi-central measures were applied. These semi-central measures were located in the green strips used for recreation and have a double use function.

Further on the retention and infiltration measures were applied on public and private properties. The private property owners were forced by regulation to implement management measures on their property, mostly realised in the recommended green strip along the property's borders.

**Figure 11.2** Aerial photo before development and infiltration strip beside road.

The drainage features were designed on the following basis:

The swale-trench-system was generally designed to cope with 1-in-5-year return period storm events and a limited outflow of 10 l/(s*ha). From history, a part of the area was designed and built already with conventional drainage. Therefore, the overall outflow of the 100-ha connected area, which is not all handled by attenuation system, was slowed down by additional semi-central swale-trench-systems to 40 l/s to improve the runoff amount to the receiving creek Wernergraben.

The objectives and challenges were met and have stood the test of time. SUDS features at Hoppegarten have been re-examined as part of a study to assess long-term performance of the technology (see Figure 11.3). Excavation of the swale was carried out to expose the geocellular storage beneath and assess sediment intrusion or blockages. Infiltration capacity was examined by a double ring permeability test and found in a good condition. Only one swale was obviously overloaded. Soil samples were taken, which showed low concentrations of pollutants, mainly found in the topsoil layer. The measures were found to be fully functional (Boogard *et al.* 2014; Sommer, 2016).

**Figure 11.3** Soil sampling and permeability test in a swale, retention measure in manhole at Hoppegarten, Germany (Sommer, 2016).

## 11.4 DISCUSSION

### 11.4.1 Biodiversity or simply effective drainage

A major achievement of the J4M8 development has been an enduring amenity value from the green infrastructure and pipe-free development on most of the site. The pipe-free network has resulted in blooms of wild flowers and amphibian and bird life. At the Amazon site, gravel filter drains were preferred to the conveyance swales of J4M8, or the mown grass under-drained swales used at Hoppegarten. For the long-term performance of the features however, the vegetated features have been found to maintain porosity, and gravel drains without plants and root activity are more likely to become silted and blocked (Bratieres *et al.* 2008; Read *et al.* 2008; Facility for Advancing Water Biofiltration [FAWB], 2009).

### 11.4.2 Keeping to a strategic plan

The J4M8 example has become a popular reference site for SUDS training in Scotland. Unfortunately, subsequent further development did not entirely follow the original drainage concept of a near pipe-free drainage system. This was perhaps due to staff changes in the planning/regulatory organisations, or policy changes that reduced time available for direct dialogue with the developers. One consequence was that conventional interceptors were preferred to grass swales or filter strips for some of the later development units, and that in turn meant deeper drains, resulting in dredging a small ditch to allow a greater depth for deep drainage pipe outfalls. The resultant lowering of the river bed released a large and continuing amount of ochre into the stream, together with associated 'marsh oil'.

Remarkably, the Amazon distribution centre in Dunfermline, although initiated some fifteen years after the initial stormwater drainage strategy document was issued, still followed the original master plan. The types of techniques used at the DEX site included a variety of source control techniques, plus a final swale/linear wetland opening into a retention pond as a final feature. That spatial application of SUDS (at source, conveyance and regional feature, as also exemplified at J4M8) corresponds to the '3 levels of treatment' set out in SEPA policy, and taken up subsequently in the original CIRIA SUDS manual, (CIRIA, 2000).

The Hoppegarten development, like that at J4M8, had to pass forward flows of stormwater from the development to a watercourse via an existing drainage pipe, requiring significant flow attenuation on site. At Hoppegarten an additional site was also developed after the main period of development. For the drainage from there to be accepted into the existing drainage, a reedbed feature was provided which attenuated pollutants and flow sufficiently to not require any modification to the drainage system downstream. It produced an additional attractive area of green infrastructure, with no environmental problems.

### 11.4.3 Risks of failures on early application of new ideas

There are a few examples of small failures at each development, sufficient to highlight risks on first time application of new technology, especially for local builders unfamiliar with the techniques. At J4M8, in some places the street edge grass swales have actually been domed rather than sloped, so that road runoff can only penetrate a short distance before either ponding, or eroding a cut through the soil and thereby down to the swale bed (Figure 11.4). It is perhaps a result of the digger driver being unaware of the requirement – probably never having built a swale previously. There are also places where the grass has grown, created new turf and bulked up sufficiently to be a barrier to sheet runoff from the road. Again this is perhaps associated with a lack of familiarity with swales; but the challenge of allowing grass turf to develop and increase in thickness at the edge of a swale is often seen to defeat swale contractors in many countries and climates.

**Figure 11.4** Some British builders were not familiar with swales; an unfortunately 'domed' section of grass swale at part of J4M8 business park.

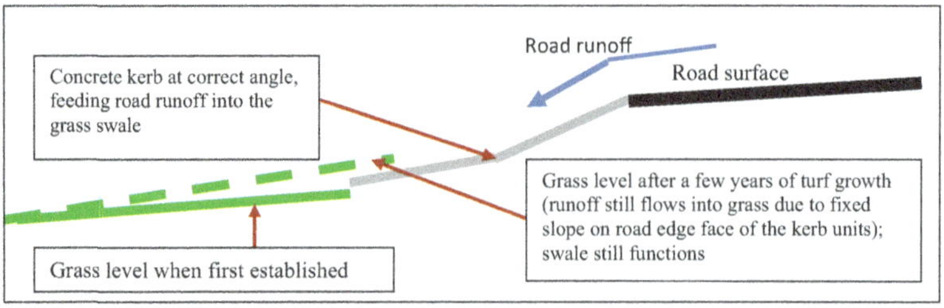


**Figure 11.5** Innovative inlet kerb designed to allow grass turf to bulk up over time and still accept inflow runoff, Hoppegarten, Germany.

Similarly at the Hoppegarten estate there are some of the best designed flat kerbs for conducting runoff from a road into a grass swale (see Figure 11.5), but these are not always used correctly.

Sadly, at Hoppegarten in one location, the excellent, innovative concrete kerbs have been laid as shallow 'V' edging along the street. That way the runoff from the road gets very close to the swale, but not into it, running parallel along the base of the 'V' instead. The majority of the swales are functional however (see Figure 11.6).

**Figure 11.6** Effective drop down kerb line along edge of road, able to feed sheet flow runoff into the under-drained grass swale (*left*) Unfortunately laid line of kerbs on one swale section illustrates risks with good innovations poorly understood by a builder (*right*) Hoppegarten 2014. *Photos credit BJ D'Arcy*

Those innovation-related failings should not have occurred at the DEX site by the time the Amazon depot was built many years after the initial master plan and start of construction on site. But nonetheless one failure has become evident: the gravel filter drain in the approach road has been simply laid with the slope of the road, rather than recognising that slope, and providing a series of flat sections with check dams and step down overflow points. Consequently some flow in wet weather from one length of the filter drain enters the drain in its upper sections – but flows out again at the lowest point, onto the road rather than continuing directly into the next part of the designed treatment train (a grass swale then a retention pond). Again the great majority of the drainage infrastructure is free from such problems. The system still functions, since the road runoff point in turn drains into a detention area in a traffic island, but the latter is at risk of early siltation by material washed out of the hundreds of metres of roadside filter drain which washes in across the road.

### 11.4.4 Multiple benefits from the SUDS features

In addition to providing cost effective drainage solutions at each case study development, the SUDS features provided a variety of ecosystem services:

- Aesthetics and well-being for staff and business visitors, associated with the green infrastructure.
- Habitat and wildlife interest – evident in the swales (e.g., frog spawn, *Rana temporaria*) and ponds (waterfowl) at J4M8 and DEX, as well as wetland native wild flowers such as marsh marigold *Caltha palustris*, cuckoo flower *Cardamine pratensis*, and marsh orchid *Dactylorhiza*. Some habitat has been also developed at Hoppegarten in the wet meadow end-of-system wetland basin area, and the reedbed.
- *Flood risk management*: an important objective achieved at each of the case study examples.
- *Pollution risk management*: ease of intervention in the swales or filter drains on approach roads within the estates as contingency planning for traffic accidents, multiple treatment interventions in treatment train of measures, and variety of treatment techniques used for day to day diffuse pollution sources.

At all the sites further engagement with the public and workforce would be worthwhile, perhaps employing interpretative signs at the key features? Additional photographs are given in Appendix 1.

#### *11.4.4.1 Innovation*

The developments contrasted in their approaches to provision of effective drainage, and each included innovative designs and ideas.

- Despite a limited problem in implementation, the great majority of the Hoppegarten swales exemplified innovation in getting road runoff effectively into the swale, with allowance for grass growth without loss of function; an important innovation.
- The J4M8 site not only achieved an almost pipe-free drainage system (also at Hoppegarten), but also created visually attractive habitat features across the development.
- *The Amazon development at DEX featured two particular innovations*: (a) a small traffic island detention basin around which delivery trucks turn whilst runoff passes into the filter media of the island; (b) the concrete road leading from the gateway into the premises is strongly built for the heavy trucks, but runoff passes down onto a permeable tarmacadam surfaced staff car park (subsequent sections of the internal perimeter road are served by their own

filter drains). The retention pond at the end of the whole development there is fed by an open swale channel which is landscaped to merge into the pond. The aesthetic, if not the wildlife interest, is spoiled however by a forbidding high steel perimeter fence.

## 11.5 CONCLUSIONS

At each of the examples, the technology utilised was cost neutral or cost-saving in comparison with conventional drainage options.

Impermeable soils did not preclude establishment of effective SUDS features and drainage.

Ecosystem services were a feature of the green infrastructure features provided in each example development.

Sustainable drainage technology was able to be built into designs for each example, where the need and other considerations could be taken into consideration at the outset.

The shallower drainage features in the SUDS used at J4M8 had no adverse impacts, but the deeper excavated stream bed for a later conventional drainage system caused pollution by ochre and marsh oil.

Care is needed and extra supervision, when new or innovative techniques are introduced to builders and associated professionals who may be unfamiliar with them. This should become less of a risk as the technology becomes more mainstream.

## REFERENCES

Boogaard F. C., Lucke T., Sommer H., De Beer J., van de Giesen N. C. and van de Ven F. H. M. (2014). Lessons Learned from Over Two Decades of Global Swale Use. 13th International Conference on Urban Drainage, 7–12 September 2014, Sarawak, Malaysia.

Bratieres K., Fletcher T. D., Deletic A. and Zinger Y. (2008). Optimisation of the treatment efficiency of biofilters; results of a large-scale laboratory study. *Water Research* **42**(14), 3930–3940.

Campbell N., D'Arcy B., Frost A., Novotny V. and Sansom A. (2004). Diffuse Pollution: An Introduction to the Problems and Solutions. IWA Publishing, London, ISBN: 1 900222 53 1.

CIRIA (2000). Sustainable urban drainage systems: A design manual for Scotland and Northern Ireland. Report C521. Construction Industry Research & Information Association, London.

CIRIA (2015). The SuDS Manual. CIRIA Construction Industry Research & Information Association, London.

D'Arcy B. J. (2013). Managing stormwater: from aspirations to routine business. International Issue 2013. *gwf-Wasser/Abwasser*, 32–38.

FAWB (2009). Adoption Guidelines for Stormwater Biofiltration Systems. Facility for Advancing Water Biofiltration. Monash University, June 2009.

Fife Council (2006). Flood Liaison and Advice Group, Guidance Note on Flooding and Drainage Issues in Relation to Planning and Development. Fife Council, Glenrothes, UK.

Horner R. R., Skupien J. J., Livingston E. H. and Shaver H. E. (1994). Fundamentals of Urban Runoff Management – Technical and Institutional Issues. Terrene Institute, Washington DC.

Read J., Wevill T., Fletcher T. D. and Deletic A. (2008). Variation among plant species in pollutant removal from stormwater in biofiltration systems. *Water Research*, **42**(4–5), 893–902.

Roesner L. A. and Campbell N. S. (1997). Dunfermline Eastern Expansion Area: Surface Water Management Study. Completion Report. Sir Frederick Snow and Partners, Edinburgh, and Camp Dresser & McKee Inc., Maitland, Florida.

Schueler T. R. (1987). Controlling Urban Runoff: A Practical Manual for Planning and Designing Urban BMPs. Metropolitan Council of Governments, Washington, DC.

Schueler T. R., Kumble P. A. and Heraty M. A. (1992). A Current Assessment of Urban Best Management Practices. Metropolitan Council of Governments, Washington, DC.

Sommer H., Kluge B., Kaiser M., Pallasch M. and Markert A. (2016). Condition and ability of long running infiltration systems. Research project funded by Ministry Climate Protection, Environment, Agriculture, Conservation and Consumer Protection of the State of North Rhine-Westphalia.

Stahre P. (2006). Sustainability in Urban Storm Drainage – Planning and Examples. Svenskt Vatten, Stockholm, Sweden. www.svensktvatten.se/sustainability

Urbonas B. and Stahre P. (1993). Storm Water Best Management Practices and Detention for Water Quality, Drainage and CSO Management. PTR Prentice Hall, Englewood Cliffs, New Jersey.

## APPENDIX: PHOTOGRAPHS OF THE THREE CASE STUDY SITES

(1) J4M8, near Livingston

Source and Site controls (filter strip and pool), feeding head of a conveyance swale, J4M8

Source control swale, J4M8, connecting with conveyance swale and retention pond

(2) Distribution depot, Dunfermline

Permeable pavement staff car park (first part of truck access road also drains into this area)

Filter drains at edge of main internal concrete road, taking runoff from truck access route, Dunfermline, Scotland, UK.

Detention basin prior to retention pond fed by linear wetland/swale and filter drain network, DEX Dunfermline, Scotland

3. Hoppegarten, near Berlin

**Source controls** at individual premises: **Permeable pavement** and **detention basins**

Combination of grassed additional car parking filter strip with grassed detention basin-swale; Hoppegarten industrial estate, near Berlin, Germany

Above: under-drained grass swale under investigation for long term effectiveness at Hoppegarten (photo credit Harald Sommer). Below: Original outlet from Hoppegarten development: still the only discharge point to small receiving watercourse.

# Chapter 12

# Maintenance requirements for stormwater management facilities

*Marla C. Maniquiz-Redillas*, Brian J. D'Arcy and Lee Hyung Kim*

**Corresponding author: marlcmani@kongju.ac.kr*

## 12.1 INTRODUCTION

The use of stormwater treatment features has increased significantly as pollution problems have been widely recognized and, in many jurisdictions, is now driven by new stormwater regulations. These regulations also require governmental units to develop a systematic approach for the documentation of inspection and maintenance, in perpetuity, to ensure that they are achieving their desired treatment goals.

There are almost as many terms in use as there are types of treatment features, but they all relate to the same technology: use of features in the built environment which can receive stormwater runoff, slow the flow, attenuate peak runoff and allow sedimentation or filtration of entrained pollutants. Some encourage infiltration after treatment, others provide for *in situ* biodegradation, but all require maintenance. Effective pollutant capture requires equally effective pollutant removal. For non-biodegradable contaminants that means physical removal by excavation or suction, or up-lifting filtration cells, etc. (and replacing them). The technology referred here encompasses all the types of system indicated in Table 12.1, which identifies example techniques and the various terms in use. An explanation of some of the terms in use for stormwater management and the associated facilities is given in Fletcher *et al.* (2015).

© IWA Publishing 2017. Wealth Creation without Pollution: Designing for Industry, Ecobusiness Parks and Industrial Estates
Brian D'Arcy, Lee-Hyung Kim, Marla Maniquiz-Redillas
doi: 10.2166/9781780408330_171

Maintenance is essential to the operation of all types of stormwater features. A poorly maintained, but still functional unit will have significantly lower pollutant removal rates. Once vegetation has died, sediment accumulation is not stabilized by vegetation, and/or erosion has become significant, the feature will remove few pollutants and may actually contribute sediments and nutrients to runoff. If not maintained, eventually features will pass sediments through – or worse, large storms could wash accumulated sediments out of the facility (Moran *et al.* 2001). In addition to maintaining the effectiveness of stormwater features and reducing the incidence of pests, proper inspection and maintenance is essential to avoid the health and safety threats inherent in some features if neglected (Skupien, 1995).

**Table 12.1** Terms for stormwater management and features to which they refer.

| Term | Techniques | Comments |
|---|---|---|
| Urban Structural Best Management Practices (Urban BMPs) | All-encompassing term; all treatment techniques, from ponds and detention basins, to biofiltration, swales, permeable pavement, and proprietary systems such as vortex separators, catchpits, interceptors, etc. | NB. BMPs are for water quality control, not flood risk management. But they can be integrated into a holistic (multi-purpose) drainage system. Spectrum of maintenance requirements. |
| Sustainable Urban Drainage Systems (SUDS). | City-derived technology, also applicable in other built environments. As above, but primarily semi-natural, passive techniques such as vegetated features, permeable surfaces and filter drains. | The integrated idea for the above: water quality and quantity, with 'amenity' (socio-economic, habitat, biodiversity, etc.) benefits too. |
| Low Impact Development (LID) | Simplistically, 'SUDS without ponds'. | Focuses on source controls to minimize land-take & allow retrofits. Maintenance may not be in public domain. |
| Proprietary BMPs | Commercially produced units, advantages are small size. Easily overlooked however on commercial/industrial premises. | Regular maintenance vital to compensate for small storage capacity. Some suppliers undertake maintenance in a lease agreement. |
| Green infrastructure drainage features | Soft-engineering BMPs/ SUDS/LID techniques, flood spill landscapes (wildflower meadows, etc. in public parks). | Especially tree planters, green roofs and other street edge or source control techniques, also large scale landscapes. Compatible with local authority or roads maintenance programmes. |

Maintenance costs for downstream features are also reduced in scale and magnitude by using the small LID practices (source controls) upstream. LID site designs require routine landscape care and maintenance of the vegetation, and regular sediment removal. This contrasts with infrequent, but potentially high costs of dredging a pond, reported at every seven years (Campbell *et al.* 2004), or longer. A study of two retention ponds in Scotland reported that sediment rates indicated about 300 years before a pond would be full, and a more practical indication of 17 years and 98 years respectively for the two ponds to reach 25% silted volume (Heal & Drain, 2003). The longer value of elapsed time to fill to 25% capacity (98 years) was a pond served by six detention basins upstream across the catchment. The more rapidly silting pond (17 years to 25%) had no such features and was an end-of-pipe facility (Heal & Drain, 2003; Jefferies *et al.* 2005). Maintenance costs are directly proportional to the field size and sediment yield. Reducing sediment yield before it reaches a pond will reduce the cost significantly. It is recommended that other sediment management practices be installed in conjunction with a sediment basin. The construction cost is a one-time cost; maintenance is a continuous cost to keep the sediment basin operating properly.

Resource guides supplement the stormwater manuals in most cities in the United States in regard to local government inspection and maintenance activities associated with various stormwater management features, and can be used as a tool by city staff and policymakers for evaluating various features to install based on anticipated long-term maintenance requirements.

## 12.2 MAINTENANCE REQUIREMENTS

Stormwater features operate at their greatest efficiency when properly maintained. In most applications, stormwater features are designed to provide a specific pollutant removal efficiency given a fully functioning system. Features are designed using a specific volume and/or surface area to settle or remove particles of sediment for a specific contributing sub-watershed. The available volume decreases over time as it fills with debris and sediment, and as such, the feature's efficiency also decreases to the point of being non-functioning if not maintained; as demonstrated in the study of Heal and Drain (2003) that measured sedimentation rates for a retention pond, extrapolating for hundreds of years to reach such a point.

In the United States, the Stormwater Equipment Manufacturers Association is a strong advocate for the proper maintenance of all types of structural Best Management Practices (BMPs) used for stormwater systems. All systems require maintenance regardless of the type of BMP installed. The maintenance requirements vary with each BMP and should be tailored according to system and site specific needs. The property owner of the BMP should be aware of the annual maintenance costs associated with each BMP and should consider these in establishing the long-term operations and maintenance plan. Table 12.2 shows a summary of suggested maintenance requirements of BMPs.

**Table 12.2** Maintenance requirements of BMPs.

| Requirement | Details |
|---|---|
| Jurisdiction | – Ensure that BMP maintenance is performed by periodic inspection of all BMPs<br>– Check the completion of required maintenance within the (agreed) time-frame<br>– Impose penalties for non-compliance |
| Inspection/maintenance service records and reports for BMPs (maintained by property owner) | – Date of inspection<br>– Name of inspection/maintenance provider<br>– Condition of BMP: proper operation; observable pollutant loads present; vegetative growth and species present; fences and other safety devices; inlet and outlet channels or structures; underground drainage, other features specific to BMP type |
| Frequency | – Inspections vary depending on BMP type; can be annual, monthly, quarterly, semi-annually, or after major storms |
| Operation and Maintenance Agreement | – Agreement is signed by the developer or BMP owner<br>– Long-term maintenance plan: include a description of the stormwater system and its components, inspection priorities, inspection schedule for each component, and a schematic for each BMP |
| Applicability | – Scheduled/regular inspection<br>– Documentation of the BMP during inspection |

*Source:* Adopted from: Stormwater Equipment Manufacturers Association, 2012.

Scheduled inspections may vary among BMPs. Certain stormwater treatment facilities may require more frequent inspections and/or maintenance than others (see Table 12.3). In order to effectively maintain stormwater facilities, cities and counties may want to specify a storm event that triggers inspection activities for these select locations. All BMPs should be inspected on a regular basis for continued effectiveness and structural integrity. During each inspection, the inspector should document whether the BMP is performing correctly, if the BMP has been damaged since the last inspection, and, if so, what should be done to repair it.

**Table 12.3** Inspection and routine maintenance frequency of selected BMPs.

| BMP Type | Inspection Frequency* | Routine Maintenance Frequency (Annual) |
|---|---|---|
| Bioretention systems | A; S | 2 |
| Cartridge or module media filtration structures | SA | 1–2 |
| Catch basin inserts (long term) | Q | 3–4 |
| Dry pond | M | 3–4 |
| Dry wells | A | 1 |
| Filter strips or swales | M | 2–3 |
| Green roofs | SA; S | 2–3 |
| Hydrodynamic or gravity separators | SA | 1–2 |
| Infiltration trenches | A; S | 2–3 |
| Permeable pavement | A | 2–3 |
| Rainwater gardens | SA; S | 2–3 |
| Rainwater harvesting | SA; S | 2–3 |
| Sand filter | Q (1st year); SA (after) | 1–2 |
| Trash and debris screens | SA; S | 2–3 |
| Underground storage facilities | SA | 1 |
| Wetlands | SA | 2 |
| Wet pond | Q | 2–3 |

*Source:* Adopted from: Stormwater Equipment Manufacturers Association, 2012.
*Note:* *A = annual; M = monthly; S = after major storms; Q = quarterly; SA = semi annually.

## 12.3 MAINTENANCE CONSIDERATIONS

The required maintenance intervals for stormwater BMPs are often dependent upon the degree of pollutant loading from a particular drainage basin. The effectiveness of post-construction stormwater control features best management practices depends upon regular inspections of the control measures. Records are important, for example documenting foul into surface water misconnections and whether effective corrective action has been taken by the appropriate agency.

Generally, maintenance can best be broken into three categories: **inspection**, **routine maintenance**, and **major maintenance**. Though each BMP type has its own unique characteristics, inspections will generally consist of an assessment to assure its functionality and the general condition. Routine maintenance is performed regularly to maintain both the aesthetics of the BMPs and their good working order. Routine maintenance will generally consist of trash and vegetation removal, unclogging of drains, minor sediment removal and exchange of filter media where applicable. Routine inspection and maintenance help prevent potential nuisances (odours, mosquitoes, weeds, etc.), and reduce the need for repair maintenance and the chance of polluting stormwater runoff by finding and fixing problems before the next rain.

Major maintenance will be completed as required from inspections and generally consists of significant reconstruction due to failures in the BMP. Examples of major maintenance include dredging, excavation, removal of existing media, replacing fabric, replacing the under-drain, and re-establishment of vegetation. It is important that routine maintenance and non-routine repair of stormwater BMPs be done according to a schedule or as soon as a problem is discovered. Because many BMPs are rendered ineffective for runoff control if not installed and maintained properly, it is essential that maintenance schedules are maintained and repairs made promptly. In fact, some cases of BMP neglect can have detrimental effects on the landscape and increase the potential for erosion. However, routine maintenance, such as mowing grasses, should be flexible enough to accommodate the fluctuations in need based on relative weather conditions. For example, more harm than good may be caused by mowing during an extremely dry period or immediately following a storm. Some examples of bad design and management of BMPs are shown in the photos below implying the need for maintenance considerations (Figures 12.1 and 12.2).

**Figure 12.1** Bad design: flow channels were not properly designed so runoff cannot be diverted into the facilities for infiltration (San Francisco/California).

**Figure 12.2** Bad management: overflow pipe in bioretention (*above left*) and small constructed wetland (*above right*) blocked with litters (San Francisco/California).

## 12.4 EXAMPLES OF ACTUAL MAINTENANCE ACTIVITIES

### 12.4.1 Small scale: Kongju National University campus, Cheonan, South Korea

LID facilities including infiltration trenches, tree box filters, subsurface flow constructed wetlands and rain garden have been developed and constructed since 2009 at the Kongju National University grounds in Cheonan City, Korea for research purposes. They are all examples of features which could be used on commercial and industrial premises. The facilities were sited either on a small landscape area near a side road or close to the edge of a parking lot to collect and treat stormwater runoff. The total catchment area designed to be treated ranges from 450 $m^2$ to a maximum of 880 $m^2$. The surface area of the facilities was typically less than 2% of the catchment area.

The maintenance activities conducted on the BMPs are shown in the photographs below (Figures 12.3–12.14). Most of the maintenance activities were focused on regular and periodic maintenance that include cleaning of the sedimentation chamber (such as removal of leaves and litters as well as accumulated sediment), inflow channel and surroundings, etc. The replacements of geotextile and the cleaning of the sedimentation zone and surroundings have contributed to the improvement of pollutant removal efficiency of the system. The maintenance is more frequent for facilities with a relatively smaller sedimentation chamber. The renovation included the cleaning of all the filter media inside the facility, and removal of accumulated sediment. Table 12.4 shows a summary of the maintenance activities conducted on the constructed facilities.

**Figure 12.3** Part of the routine maintenance performed in the tree box filter was the inspection of possible accumulation of large debris and litters (*above left*) and checking of standing water and possible clogging (*above right*).

**Figure 12.4** Soil settlement (*above left*) and geotextile clogging (*above right*) were also inspected as part of routine maintenance for the infiltration trench.

**Figure 12.5** Sediment, debris and litter deposition in the sedimentation and facility outlet to drainage access were inspected as part of routine maintenance for the hybrid constructed wetland.

**Figure 12.6** After inspection, litters and debris were collected as part of routine maintenance.

Before After

**Figure 12.7** Another routine maintenance performed for the hybrid constructed wetland was the management of vegetation and ground cover conducted during the plant dormancy period in winter.

**Figure 12.8** Sediments accumulated in the pre-treatment tank of the infiltration trench (*above left*) and overflow area of the hybrid constructed wetland (*above right*) were removed as part of the non-routine maintenance.

Before After

**Figure 12.9** Woodchip in the vertical media cartridge was replaced as part of the non-routine maintenance in the infiltration trench.

**Figure 12.10** Filter media were collected and washed as part of the non-routine maintenance.

**Figure 12.11** As part of the non-routine maintenance, geotextile fabric was replaced in the pretreatment tank of the infiltration trench (*above left*) and hybrid constructed wetland (*above right*).

**Before** **After**

**Figure 12.12** The inflow channel of the infiltration trench was renovated as part of the non-routine maintenance.

Before After

**Figure 12.13** Filter media were packed for ease of maintenance in the infiltration trench as part of the non-routine maintenance.

Before After

**Figure 12.14** A flow diversion channel was constructed to ensure flow diversion to the hybrid constructed wetland as part of the non-routine maintenance.

## 12.4.2 Medium scale: Sejong City, Korea

The Multifunctional Administrative City (MAC) in Sejong, Korea, which has been under development since 2007 with completion due in 2030, will be an internationally recognized planned city. The new city was conceived to eliminate urban sprawl in the capital region and promote balanced regional economic development. Sejong will accommodate self-sufficient functions such as administration, culture, education, research, medical care, welfare, and high-tech industries. With vast green and water spaces reaching more than half of the city area and a forward-thinking energy policy, MAC will be one of the world's greenest cities.

The character of Sejong is in its sustainable development plan to improve the long-term social and ecological health of the city. Concurrent to the efficient land use and living environments and a sustainable economy, low impact development has also been incorporated in Sejong. Table 12.5 shows the characteristics of some

**Table 12.4** Suggested maintenance activities applicable at the constructed small to medium scale BMP facilities.

| Routine Maintenance | Procedure/Details |
|---|---|
| Visual inspection | – Check the inlet structure for large debris/litters that may block the passage of runoff during a storm<br>– Check for misalignment of the structure or possible settlement of the surrounding soil (different from the original design)<br>– Check for standing water<br>– Check for sediment deposition<br>– Check for clogging in the geotextile<br>– Check the outlet structure<br>– Check the access to the upstream and downstream drainage |
| Cleaning or removal of litters and debris | – Sweeping of litters, leaves and debris at the pretreatment tank, outflow tank, and overflow tank<br>– Sweeping of large particulates at the inlet & outlet structure, gutter or curb of the facility |
| Vegetation and ground cover management | – Check for the percentage of vegetation cover<br>– Check if the vegetation needs to be replaced |
| **Non-routine Maintenance** | **Procedure/Details** |
| Cleanout trash and accumulated solids | – The sediment accumulated at the pretreatment tank, outflow and overflow areas should be removed |
| Washing and replacement of media filters | – Necessary if apparent sediment accumulation is observed or if the runoff is not flowing well |
| Geotextile fabric | – Geotextile fabric should also be replaced whenever sediment deposition is too great |
| Structural and other repairs | – Flow path should be reconstructed if runoff is not properly entering the facility<br>– Metal grates (covers) should be changed if corrosion is evident (might affect the metal concentration of runoff)<br>– Other designs such as the chains, fence, and educational board should be repaired if unwanted damage has occurred |
| **Major Maintenance** | **Procedure/Details** |
| Renovation | – Necessary if the measured removal percentages are very low or if the volume reduction achieved is not satisfactory<br>– Necessary if excessive erosion or channelization occurs that changes the original design and slope of the system |
| Rebuild | – Necessary if the drainage system malfunction or the soil is totally saturated affecting the groundwater table |

examples of BMPs installed in Sejong city with actual site photographs of regularly maintained BMPs (Figure 12.15). The typical maintenance activities applied are summarized in Table 12.6.

**Table 12.5** Characteristics of LID facilities installed in Sejong city.

| BMP Type | Parameter | | | | |
|---|---|---|---|---|---|
| | Catchment Area ($m^2$) | Land Use (cover) | Design Rainfall (mm) | WQV ($m^3$) | Dimension (W*L*H, m) |
| Infiltration trench | 1620 | Road (Asphalt) | 5.3 | 8.2 | 2.65 × 31.25 × 1.65 |
| Rain garden | 2340 | Road (Asphalt) | 5.3 | 11.8 | 2.60 × 26.20 × 1.55 |
| HSSF constructed wetland | 3600 | Bridge (Concrete) | 5.3 | 18.1 | 6.30 × 15.70 × 1.65 |
| FWS constructed wetland | 3200 | Bridge (Concrete) | 5.3 | 16.1 | 5.20 × 23.10 × 1.70 |
| Vegetated swale | 1020 | Road (Asphalt) | 5.3 | 10.8 | 2.90 × 19.60 × 1.85 |
| Small constructed wetland | 950 | Road (Asphalt) | 5.3 | 4.8 | 2.80 × 8.60 × 1.55 |

WQV: water quality volume, HSSF: horizontal subsurface flow, FWS: free-water surface.

**Figure 12.15** On-site photographs of newly-constructed and regularly maintained BMPs in Sejong city, South Korea.

**Table 12.6** Maintenance activities applicable to medium and large scale BMPs.

| BMP Type | Maintenance Activities |
|---|---|
| Retention pond/basin | – Removal of litters and dredging sediments from inlet/outlet channels<br>– Measurement of the sediment accumulation in pretreatment and retention basins and dredging the sediments |
| Constructed wetland | – Removal of litters and dredging sediments from inlet/outlet channels<br>– Measurement of the sediment accumulation in sedimentation basin of wetland<br>– Management of the vegetation for appropriate plant cover and growing<br>– Removal of the invaded species and some of the dead plants (left more than 10 cm above water surface)<br>– Management of the landscape by mowing<br>– Check and repair of the facility<br>– Maintain the plant cover over more than 40% of water surface in order to reduce the eutrophication and to provide dissolved oxygen |
| Infiltration facility | – Removal the debris from surface and cover of infiltration facility<br>– Determination of the frequency of media changes by checking the infiltration rate<br>– Check and manage the differential settlement<br>– Check differential settlement, erosion, and clogging<br>– Check the groundwater table during dry periods<br>– Vegetation control in the facility |
| Vegetated filter strips | – Removal of litters and dredging sediments from inlet/outlet channels<br>– Plant management during summer and removal of dead plants during winter<br>– Check the soil settlement or compaction under the plants |
| Filtration facility | – Removal of litters and dredging sediments from inlet/outlet channels<br>– Check for clogging and replacement of media<br>– Check the sediment accumulation rate above filters |

## 12.4.3 Medium to large scale

### *12.4.3.1 Four rivers' BMP demo projects, South Korea*

The Four River Restoration Project envisioned having additional water sources for South Korea. During this project however, it was identified that although water quality discharge from point source pollution facilities were already controlled, water quality in the four major rivers of South Korea had not been improving. This finding led to the addition of stormwater management programmes such as Four Rivers' BMP demo projects to control nonpoint source pollution thereby protecting natural water bodies. At present, installation of LID and BMP technologies have been added as per the legal requirement for all industrial complexes in South Korea. The functions of these LID technologies in managing stormwater runoff from industrial areas may be optimized by performing the following maintenance activities (Figure 12.16 (a)–(d)):

The acceptability or otherwise of uncut vegetation (essential for overwintering invertebrates and as shelter for other wildlife) is a matter for education. It has become good practice in many countries to only mow enough grass around a feature (e.g., one section for easier access) to show that the feature is being maintained, not merely overlooked. It is of course also a significant cost saving for larger features. Removal of dead plants in winter may not need to be 100% of the area on each occasion. There's a balance between excessive cost incurred by extreme cutting, and the acceptability to owners and their customers of keeping some winter habitat.

**Figure 12.16** Some examples of maintenance activities four major rivers' BMP sites in South Korea. (a) Dredging for sediment removal, (b) Litter removal, (c) Gardening for plant removal and (d) Removal of dead plants

Table 12.7 shows maintenance costs, specific activities, and schedules for several post-construction runoff BMPs in Korea. Solid wastes refer to litter, debris and sediments. In addition to those disposal costs and labour costs for all the maintenance work, there are other costs which include rental costs for equipment to be used for maintenance procedures. Although Table 12.7 reports a lot of information for non-industrial commercial features, the general scale of maintenance requirements and associated costs for BMPs in Korea is useful to compare with other examples. On industrial premises source control measures (housekeeping, such as bunds around oil storage tanks, and equivalent appropriate measures for chemicals and other potential pollutants), as well as stormwater features to manage runoff, are all needed if maintenance costs of downstream final controls (such as ponds or extended detention features) are to be minimized. If the tonnes of waste removed during maintenance of facilities in Table 12.7 are contaminated with oil or solvents for example, disposal costs will be very high.

### *12.4.3.2 Examples in the USA and UK*

The examples shown below (Figures 12.17–12.19) explain why maintenance costs of stormwater management facilities can be so high. In the USA grass lawns are favoured landscapes, and require frequent mowing. They also require fertilizer to give the optimum appearance for the grass lawn enthusiasts. Experience in many cities in North America (and in places in Europe too) is that such extensive short grass areas are also attractive to nuisance populations of feral Canada geese. The waterbody consequently is liable to receive excessive quantities of nutrients, leading to algal blooms and anaerobic conditions in summer. The fountain in the example below is likely to be an aeration unit as much as a water feature – and it is another operational cost. In several North American cities expensive herbicides are a major cost, amounting to the majority of the costs in some situations (D'Arcy, 1998). The communities in such examples can save money on maintenance by moving towards habitat creation, especially alongside the water margins, to reduce the load of avian faecal material in runoff, to reduce the area of grass to be mown, and to reduce the quantities of soluble nitrate and phosphate added and likely to be mobilized in runoff. The nutrients cause the blooms of algae and pondweeds which in turn incur the costs for poisoning and removal.

On industrial premises the risks of significant oil or other degradable materials being spilled requires care when disturbing sediments since they may be anaerobic and a source of unpleasant odour when brought to the surface (Wilson *et al.* 2003). A good sense of smell is an asset when inspecting industrial stormwater features; detection of significant spills can allow material to be removed prior to problems developing.

**Table 12.7** Frequency and maintenance cost of constructed BMPs in Korea.

| BMP | Location | Catchment Area (ha) | Facility Volume ($m^3$) | Facility Area ($m^2$) | Land-Use | Yearly Frequency of Maintenance | | | Yearly Cost (Korean Won)* | | |
|---|---|---|---|---|---|---|---|---|---|---|---|
| | | | | | | Dredging of Sediment | Removal of Plants/ Landscaping | Inspection and Others | Labour Cost | Solid Wastes | Others |
| Infiltration trench | Cheonan | 0.052 | 6.5 | 5 | road | 1 | 2 | 6 | 210,000 | 200,000 | 100,000 |
| | Yongin | 0.3 | 39 | 300 | road | 1 | 2 | 12 | 1,680,000 | 2,500,000 | 630,000 |
| | Yongin | 0.5 | 55 | 300 | road | 1 | 2 | 12 | 1,680,000 | 2,500,000 | 630,000 |
| Tree box filter | Cheonan | 0.088 | 2.9 | 2.25 | parking lot | 1 | 1 | 6 | 120,000 | 100,000 | 100,000 |
| | Cheonan | 0.045 | 2.93 | 2.25 | parking lot | 1 | 1 | 6 | 135,000 | 100,000 | 100,000 |
| | Suwon | 3.4 | 174 | 62 | road | 1 | 2 | 12 | 1,680,000 | 792,000 | 380,000 |
| Small hybrid wetland | Cheonan | 0.0597 | 4.55 | 6.5 | road/ parking lot | 1 | 1 | 6 | 135,000 | 100,000 | 100,000 |
| | Cheonan | 0.0457 | 4.9 | 7 | road/ parking lot | 1 | 1 | 6 | 135,000 | 100,000 | 100,000 |
| Rain garden | Cheonan | 0.02 | 16.2 | 10.8 | road | 1 | 2 | 6 | 210,000 | 200,000 | 100,000 |
| Bioretention | Cheonan | 0.0139 | 3.6 | 3 | parking lot | 1 | 2 | 6 | 180,000 | 200,000 | 100,000 |
| | KEC, Incheon | 4.43 | 709.6 | 3,365 | road/ urban | 1 | 2 | 12 | 1,680,000 | 2,300,000 | 600,000 |
| Planter | Cheonan | 0.0813 | 2.4 | 1.6 | roof | 1 | 2 | 6 | 180,000 | 200,000 | 100,000 |
| Vegetated swale | Yongin | 0.77 | 40.8 | 346 | road | 1 | 2 | 12 | 1,680,000 | 2,500,000 | 630,000 |
| | Yongin | 27.66 | 420 | 1,708 | road/agro/ forest | 1 | 2 | 12 | 1,680,000 | 2,500,000 | 630,000 |
| Vegetated strips | Yongin | 2.53 | 84 | 1,966 | dry paddy field/forest | 1 | 3 | 12 | 1,800,000 | 2,500,000 | 650,000 |
| | Yongin | 7.03 | 98 | 3,258 | agro | 1 | 3 | 12 | 1,800,000 | 2,500,000 | 650,000 |

(*Continued*)

**Table 12.7** Frequency and maintenance cost of constructed BMPs in Korea (*Continued*).

| BMP | Location | Catchment Area (ha) | Facility Volume ($m^3$) | Facility Area ($m^2$) | Land-Use | Yearly Frequency of Maintenance | | | Yearly Cost (Korean Won)* | | |
|---|---|---|---|---|---|---|---|---|---|---|---|
| | | | | | | Dredging of Sediment | Removal of Plants/ Landscaping | Inspection and Others | Labour Cost | Solid Wastes | Others |
| FWS constructed wetland | Yongin | 10.38 | 893 | 4181 | agro | 1 | 4 | 12 | 1,920,000 | 3,500,000 | 820,000 |
| | Icheon | 7.42 | 1741 | 5010 | road/agro/ urban | 1 | 4 | 12 | 1,920,000 | 3,500,000 | 820,000 |
| | Namyangju | 326 | 16,300 | 1000 | urban/ agro | 1 | 4 | 12 | 1,920,000 | 3,500,000 | 820,000 |
| | Nonsan | 11 | 4006 | 9778 | livestock area | 1 | 4 | 12 | 1,920,000 | 3,500,000 | 820,000 |
| | Jungup | 60 | 3799 | 9399 | livestock area | 1 | 4 | 12 | 1,920,000 | 3,500,000 | 820,000 |
| | Kongju | 465 | 2957 | 8943 | livestock area | 1 | 4 | 12 | 1,920,000 | 3,500,000 | 820,000 |
| | Kongju | 221 | 11,235 | 23772 | agro | 1 | 4 | 12 | 1,920,000 | 3,500,000 | 820,000 |
| | Gimje | 75 | 1836 | 7804 | agro | 1 | 4 | 12 | 1,920,000 | 3,500,000 | 820,000 |
| Infiltration-retention basin | Yongin | 9.13 | 517 | 3325 | road/ forest | 1 | 2 | 12 | 1,680,000 | 3,500,000 | 780,000 |
| | Yongin | 9.61 | 500 | 3463 | industrial/ road/ housing/ agro | 1 | 2 | 12 | 1,680,000 | 3,500,000 | 780,000 |
| Retention basin | Kwangju | 177.3 | 9162 | 9160 | road/ housing/ forest | 1 | 2 | 12 | 1,680,000 | 3,500,000 | 780,000 |
| | Kwangju | 1.6 | 170 | 103 | road/ housing | 1 | 2 | 12 | 1,680,000 | 3,500,000 | 780,000 |

*1 US$ = 1100 KRW.

**Figure 12.17** Maximizing maintenance costs: conventional design and management of (a) a retention pond (*above left*) and (b) pond and vegetated strips (*above right*) for treating runoff from industrial areas (Cincinnati).

**Figure 12.18** Good design and management of constructed wetlands in industrial areas (Cincinnati), from a cost-saving maintenance perspective as well as for wildlife. Similar maintenance choices are evident in the smaller features below.

**Figure 12.19** Industrial estate wetlands in Kinross, Scotland, UK (photo credits BJ D'Arcy).

## 12.5 DISCUSSION

As noted above in USA examples, mowing of vegetated and grassed areas may be the costliest routine maintenance consideration (Water Environment Federation [WEF], 1998). Provision of stormwater features on industrial and commercial premises at the rear of the buildings is less dominated by aesthetic considerations; out of site of the public there is plenty of scope for reducing costly aesthetic elements of conventional maintenance programmes. But the grass lawn and tidy image is just as important at the front of the office as anywhere else. Short grass will show misdemeanours such as casual disposal of oil and chemicals where runoff is across a grass strip rather than into a drain, and that will facilitate management action to control such actions, so it has practical benefits where a surface drainage system rather than a piped system is used.

Measures at source (housekeeping and stormwater management) should also help management at individual premises capture potential pollutants for re-use or recycling. Economically significant losses of oil and other raw materials and products can occur over time from leaking glands on transfer pumps; minor fractures in underground pipes, for example transferring from a storage tank to a production unit; and oil or chemical residues dripping from scores of ‘empty’ drums in a yard (D’Arcy & Bayes, 1995). With concealed (underground) conventional drainage systems such failures can escape detection for long periods, but an inspection and maintenance regime for stormwater features (as well as housekeeping/source risk areas) can allow management to detect problems more speedily and address them. An effective maintenance regime at a factory, which highlights oil contamination, for example staining the vegetation in a swale or biofiltration unit, can encompass all those aspects and be an asset for site managers.

## 12.6 CONCLUSION

There is an inescapable need for stormwater management on individual industrial and commercial premises, both as part of addressing the flood risks associated with extensive impervious surfaces, and to provide source control measures to capture contaminants in runoff. Stormwater management features will need maintenance, as everywhere. On industrial/commercial premises, an inspection and maintenance regime will also be of importance for management to be fully aware of operational practices in relation to casual disposal of material, and initially minor failures which could become more significant if not quickly addressed.

The more stringent stormwater regulations have contributed to the increase in the use of stormwater treatment strategies and management features. The regulations also require the development of a systematic approach for the documentation and implementation of inspection and maintenance of these features. BMPs operate at their greatest efficiency when properly maintained. Thus, ongoing inspection and maintenance are essential to ensure that BMPs are achieving their desired treatment goals.

In comparison with the maintenance costs for traditional stormwater controls, many of the integrated stormwater management features require little more than normal landscaping maintenance treatment. Since (routine and non-routine) maintenance work is usually not technically complicated, workers can be drawn from a large labour pool. But, as structural stormwater management features increase in their sophistication, however, more specialized maintenance training might be needed to sustain their effectiveness. The use of pretreatment and other sediment management practices will also help to reduce the maintenance burden, since maintenance costs are directly proportional to the field size and sediment yield. The construction cost is a one-time cost; maintenance is a continuous cost to keep the systems operating properly.

## REFERENCES

Campbell N., D'Arcy B., Frost A., Novotny V. and Sansom A. (2004). Diffuse Pollution: An Introduction to the Problems and Solutions. IWA Publishing, London. ISBN: 1 900222 53 1.

Center for Watershed Protection (CWP) (1998). Costs and Benefits of Stormwater BMP's. Final Report 9/14/98. Center for Watershed Protection, Ellicott City, MD.

D'Arcy B. J. (1998). Unpublished technical report, SEPA East Region.

D'Arcy B. J. and Bayes C. D. (1995). Industrial Estates: A Problem. In: C. Pratt (ed.), Proceedings of the Tenth Meeting of the Standing Conference on Stormwater Source Control, School of the Built Environment, Coventry University, Coventry. ISBN 0 905949 33 1.

Fletcher T., Schuster W., Hunt W. F., Ashley R., Butler D., Arthur S., Trowsdale S., Barraud S., Semandini-Davies, Bertrand-Krajewski J.-L., Mikkelsen P. S., Rivard G., Uhi M., Dagenais D. and Viklander M. (2014). SUDS, LID, WSUD and more – The evolution and application of terminology surrounding urban drainage. *Urban Water Journal*, **2014**, doi: 10.1080/1573062X.2014.916314.

Heal K. and Drain S. (2003). Sedimentation and sediment quality in ponds. In: SUDS in Scotland – The Monitoring Programme, C. Jefferies (ed.), SNIFFER (Scotland & Northern Ireland Forum for Environmental Research), pp. 64–68, Final Report, SR SR(02)51, August 2003.

Jefferies C., Duffy A., Spitzer A., Heal K. and Lancaster J. (2005). MUDTWADE, contract report. Abertay University, Dundee, UK.

Montgomery County Department of Environmental Protection (1999). Maintaining Urban Stormwater Facilities: A Guidebook for Common Ownership Communities. http://www6.montgomerycountymd.gov/ocptmpl.asp?url=/content/ocp/ccoc/ccoc_index.asp

Santa Clara Valley Urban Runoff Pollution Prevention Program (2005). Example BMP Inspection and Maintenance Checklist. www.scvurppp-w2k.com/bmp_om_forms.htm

Skupien J. (1995). Post-construction Responsibilities for Effective Performance of Best Management Practices. In: National Conference on Urban Runoff Management: Enhancing Urban Watershed Management at the Local, County, and State Levels. Seminar Publication. EPA 625-R-95–003. U.S. Environmental Protection Agency, Office of Water, Washington, DC.

Stormwater Equipment Manufacturers Association (2012). Inspection and routine maintenance frequency of selected BMPs. http://www.stormwaterassociation.com/

USEPA (2000). Fact Sheet 2.6: Stormwater Phase II Final Rule, Construction Site Runoff Control Minimum Control Measure. EPA 833-F-00–008. U.S. Environmental Protection Agency, Office of Water, Washington, DC.

Water Environment Federation (1998). Urban Runoff Quality Management. WEF Manual of Practice No. 23, ASCE Manual and Report on Engineering Practice No. 87. Water Environment Federation and American Society of Civil Engineers, Alexandria, VA.

Wilson S., Bray R. and Cooper P. (2003). Sustainable Drainage Techniques. Hydraulic, Structural and Water Quality Issues. Construction Industry Research and Information Association, CIRIA, London, pp. 111–115.

# Chapter 13

# The Interaction between the EU Industrial Emissions and Water Framework Directives with particular emphasis on industrial estates

*Andrew Farmer*

*The Institute for European Environmental Policy (Email: afarmer@ieep.eu)*

## 13.1 INTRODUCTION

Industrial estates are a complex dilemma for regulators, but they also present an opportunity for integrated and creative thinking to optimise environmental protection choices, minimise administrative costs and maximise industrial synergies. This paper is effectively a case study of the European Union (EU) examining the constraints and opportunities for regulation under the EU's Industrial Emissions Directive (IED) (EU, 2010) and, alongside that, the need to meet the objectives of the EU Water Framework Directive (WFD) (EU, 2000). Both of these directives have specific details which are unique to those laws and there are a range of legal and practical interactions in their implementation (Farmer & Cherrier, 2010, 2011). However, there are lessons for those in countries outside of the EU of wider interest.

This paper firstly examines the regulatory approach of these two key pieces of legislation relevant to industrial estates. It then considers the nature of the interaction between these directives before considering the opportunities and constraints to integrated approaches to regulation of industrial estates more widely.

## 13.2 THE INDUSTRIAL EMISSIONS DIRECTIVE

The IED requires specific activities to operate to conditions set out in permits issued by regulators. These conditions should include, inter alia, emission limit values based on Best Available Techniques (BAT). The 'unit' of regulation of the

© IWA Publishing 2017. Wealth Creation without Pollution: Designing for Industry, Ecobusiness Parks and Industrial Estates
Brian D'Arcy, Lee-Hyung Kim, Marla Maniquiz-Redillas
doi: 10.2166/9781780408330_193

IED is the 'installation'. It is the understanding of what 'installation' means that is central to the scope of regulation under IED and what is or is not included with respect to industrial estates.

The IED defines an 'installation' (Article 2.1.3) as 'a stationary technical unit within which one or more activities [...] are carried out, and any other directly associated activities on the same site which have a technical connection with the activities [...] and which could have an effect on emissions and pollution'. With regard to the IED industrial estates may:

- Consist of several IED installations.
- Consist of a mixture of IED installations and non-IED installations.
- Contain activities 'directly associated' with non-IED installations.
- Include other relationships between the installations/activities on the estate.

Further, Article 4.2.2 states that Member States may opt to provide that a permit cover two or more installations or parts of installations operated by the same operator on the same site. Such a permit shall contain conditions to ensure that each installation complies with IED requirements.

The definition of 'installation' and of what is a 'directly associated' activity has resulted in much debate and the European Commission (EC, 2007) has issued guidance on this issue. The guidance makes clear that activities are only included in the installation if they are 'directly' associated and 'technically connected'. If there is no technical connection, for example an administrative office, they are not included. Further, to be included 'the operation of the non-Annex I activity must somehow be closely related with the Annex I activity in a direct operational sense'. 'Technical' is 'interpreted to mean that there is a link in terms of intended process operation and materials flow. For instance, two activities can be said to be technically connected if they are operated as part of what can reasonably be viewed as a single overall operation, even if the nature of the connection is by means other than a permanent physical link'. However, while a power station supplies electricity to local activities and this is a technical connection, this would not be sufficient to be considered to be captured by the term 'installation'.

The guidance provides some examples of 'general types of non-Annex I activities' that may be directly associated with and technically connected to Annex I activities. Each of these may also produce emissions to water. They are:

- 'combustion units that provide heat and/or power;
- activities for the supply, handling and preparation of raw materials used as process inputs;
- activities concerned with the handling of intermediate products (e.g., where there are two Annex I activities and an intermediate activity between them);
- activities concerned with the handling (e.g., finishing, storage) of products; and
- activities concerned with the treatment or storage of by-products, wastes or emissions (e.g., effluent treatment units).'

However, overall the guidance notes that in some cases the decision is a matter of judgement. The guidance suggests that if much of the activity of the non-IED activity is associated with an IED activity, then it should be considered to be directly associated, but if most of its activity is associated with other non-IED activities, it could be reasonably considered to be non-associated. The guidance further states that if a wastewater treatment facility serves several facilities, one of which is under the IED, if that treatment facility is considered to be directly associated, this does not mean that the non-IED facilities which it also serves are also directly associated.

## 13.3 REQUIREMENTS FOR MEASURES TO BE ADOPTED UNDER THE WATER FRAMEWORK DIRECTIVE

The Water Framework Directive applies to surface freshwaters, groundwaters and coastal marine waters. The purpose of the WFD (Article 1) is to establish a framework for the protection of surface and ground waters which, inter alia:

- prevents further deterioration and protects and enhances the status of aquatic ecosystems;
- aims at enhanced protection and improvement of the aquatic environment, inter alia, through specific measures for the progressive reduction of discharges; and
- ensures the progressive reduction of pollution of groundwater and prevents its further pollution.

This is further elaborated in Article 4, which requires Member States to prevent deterioration of ecological quality and pollution of surface waters and restore polluted waters, in order to achieve good ecological status in all surface waters by 31 December 2015 (subject to potential delays for two further River Basin planning cycles – 2021 & 2027). Good Status is defined variously for surface and ground waters according to detailed criteria based on biology, chemistry and hydromorphology. Discharges from IED installations may affect chemical status (of surface or ground waters), or directly affect biological status (e.g., via thermal discharges). Thus an IED installation might affect the achievement of Good Status through affecting different elements that comprise Good Status. In order to achieve the objectives of the WFD, Member States are required to develop a River Basin Management Plan (RBMP) which sets out the objectives and pressures in a river basin and contains a Programme of Measures to address the pressures and so meet the objectives of the WFD.

Article 11 of the WFD defines measures for inclusion in programmes of measures within RBMPs. These are divided into 'basic' and 'supplementary' measures. Programmes of measures 'shall include' basic measures, they are 'the minimum requirements to be complied with' and they 'shall consist' of a series of different types. Art. 11.3 lists the different types of 'basic measures'. Some authorities or Member States have focused their attention on Art. 11.3.a – measures that are required by other EU laws (e.g., the Nitrates Directive). However, Art. 11.3

lists a further nine types of basic measures which are also minimum requirements. With regard to industrial estates, the following are most relevant:

(a) for point source discharges liable to cause pollution, a requirement for prior regulation, such as a prohibition on the entry of pollutants into water, or for prior authorisation, or registration based on general binding rules, laying down emission controls for the pollutants concerned [….] These controls shall be periodically reviewed and, where necessary, updated;
(b) for diffuse sources liable to cause pollution, measures to prevent or control the input of pollutants. Controls may take the form of a requirement for prior regulation, such as a prohibition on the entry of pollutants into water, prior authorisation or registration based on general binding rules where such a requirement is not otherwise provided for under Community legislation. These controls shall be periodically reviewed and, where necessary, updated;
(c) a prohibition of direct discharges of pollutants into groundwater.

Beyond the basic measures, the WFD states (Art. 11.4) that 'supplementary measures' may also be adopted 'in addition to the basic measures' to meet the obligatory objectives of the WFD and also 'to provide for additional protection or improvement' of water bodies. Further, Article 10 requires the adoption of the 'combined approach' for point and diffuse sources of pollution, including:

- The emission controls based on best available techniques, or
- The relevant emission limit values, or
- In the case of diffuse impacts the controls including, as appropriate, best environmental practices.

Industrial estates have the potential to result in both point and diffuse sources of pollution to water. The WFD does not address the issue of the way in which appropriate controls are to be established. For example, a regulator could adopt a series of different measures for separate activities on an industrial estate. Alternatively, a regulator may take an integrated approach. Ultimately the WFD requires an understanding of the pressures arising from an industrial estate and it is up to the regulator to determine the measure or suit of measures necessary to address that pressure. The requirements of existing EU law are a first, but not sufficient, starting point. Other basic measures are required.

A key area of concern for water pollution and industrial estates is diffuse runoff to surface or ground waters. Point source emissions can be more easily regulated and may be brought together in common wastewater treatment plants or addressed via sewer controls. Indeed, they may become captured within the regulatory framework of the IED. Diffuse pollution is harder to capture in a regulatory framework with multiple potential sources and operators. However, a coherent control approach is clearly beneficial – taking action to avoid runoff, ensuring surfaces are hard and do not allow seepage of pollutants into soils, etc. If a regulator has concerns that accidents may occur on an estate, then some common approaches to controlling such pollution may be in place. In any case, the WFD allows for any measure to be implemented which meets the objectives of the directive. Further, the most cost effective measures are most desirable. For this reason, an estate-wide approach to managing pollutant discharges/spills and containing and treating these is desirable. The only limitation is whether there are constraints in a country to taking such an approach.

## 13.4 INTERACTION BETWEEN THE IED AND WFD

The EU Network for the Implementation and Enforcement of Environmental Law (IMPEL) has undertaken two studies on the interactions between these directives – the first a legal/policy analysis (Farmer & Cherrier, 2010) and the second a collation and analysis of experience from Member State IED regulators and WFD water managers (Farmer & Cherrier, 2011). The first study identified a number of challenges that the interactions between the directives pose to the competent authorities of the Member States and how these might be addressed. The range of different types of interactions is illustrated in Figure 13.1.

There may be legal uncertainty, for example due to inconsistencies between Directives and Regulations. In most cases there is consistency between the legislation, but there may be different national interpretations of obligations which may result in unintended barriers to integration of the implementation of the Directives.

Member States interpret the scope of interpretation of the IED differently, i.e. what is included within a permit. Deciding what is included within IED regulation can assist in helping to deliver the objectives of EU water Directives. This particularly concerns directly associated activities (such as wastewater treatment, manure spreading, etc.).

The WFD and IED ‘management units’ are at different spatial scales. The WFD is focused on the scale of the river basin (or water body), while the IED focuses on the installation. This presents challenges for integration between them. In particular the spatial, landscape approach to river basin management can be a different thought process to site-based analysis under the IED. Defining obligations on IED installations is a challenge – how to translate understanding of pressures from installations on water objectives under EU water law into discharge requirements for permits.

The IED includes a new obligation to consider environmental issues in enforcement activity. This is a new obligation that will require inspectorates not only to consider whether permits are complied with, but also to examine impacts on the local environment, providing a greater link to examining relationships between IED installations and water objectives.

A problem arises from the fact that the Directives have been implemented over non-complimentary timetables. IED permits may have been issued before water objectives are defined. Revisiting them may impose costs, but there are concerns over whether some are IED compliant. The WFD may provide added impetus to address any implementation deficiencies.

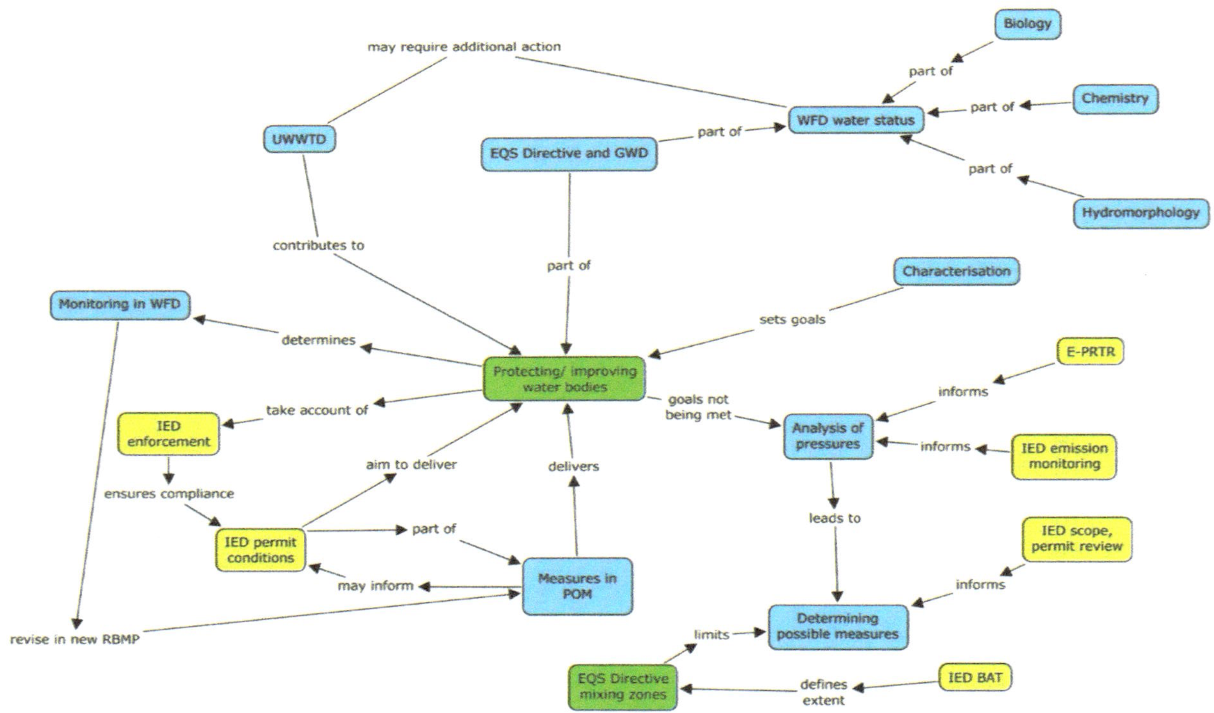


**Figure 13.1** The interactions between EU law relating to industrial pollution control and water management.
Note that boxes in blue are largely the responsibility of water management authorities, those in yellow the responsibility of IED competent authorities and those in green are a joint responsibility (EQS environmental quality standard; GWD groundwater directive; UWWTD urban waste water treatment directive; POP priority organic pollutants).

The Directives have their own monitoring obligations and integrating these with the need for information transfer between different authorities requires close collaboration between authorities. Also systems need to be in place to ensure full information transfer in ways that are sufficient to support implementation of the relevant legislation.

Under the IED, the Commission and Member States have exchanged information to produce Best Available Techniques reference documents (BREFs). However, these currently provide little guidance in relation to water objectives arising from EU water law to help in decisions on individual installations. Revision of the BREFs under the IED provides an opportunity to rectify this.

In order to support Member States in the implementation of the WFD, a wide range of guidance documents have been produced under the Common Implementation Strategy (CIS). This guidance provides a large amount of information to support the WFD, but consideration of the relationship with the IED is often limited. This may be an issue to be addressed as guidance is revised in the future.

The second IMPEL study found that there is significant complexity with multiple sources of pollutants to water (IED and/or non-IED), which is a regulatory challenge for industrial regulators and water authorities. They need to accurately assess the relative importance of the different sources regarding pressures of concern. IED permit conditions need to ensure installations operate so as not to threaten the objectives of the water Directives which may require going 'beyond' BAT.

The institutional relationships between the IED and water authorities vary enormously between Member States. It is important to put procedures in place to facilitate ways of working together to ensure that the right information is shared, that information exchange is timely and that management decisions are, therefore, are more robust. Co-ordination and co-operation are key factors for success.

## 13.5 INTEGRATED REGULATION FOR AN INDUSTRIAL ESTATE

Large industrial estates with multiple operators provide opportunities to make use of synergy effects among different activities. The regulator may take the overall environmental impact of the combined activities on an estate into consideration when issuing individual permits for these activities. Permits could oblige operators to take into account the synergistic effect with/on neighbours as well as ensuring the requirements of the WFD are complied with. Such a permit may be possible under the IED or it could be developed within a national regulatory framework. In some cases it may not be possible and it can have disadvantages. The advantages of a single permit would be:

- Single permit and documentation, reducing administration.
- A truly integrated approach assessing impacts on the environment as a whole, including assessing interactions between activities.

- Ability to assess interactions between different activities.
- Optimisation of pollution control and monitoring systems.
- Integrated wastewater management where appropriate.
- Cost-benefits to industry due both to administrative integration and achieving cost-effective measures on a large scale.
- Cost-effectiveness and environmental benefits possible.

The disadvantages of a single permit would be:

- Probable lack of clear responsibilities for individual operators.
- Where such permits are not obligatory, it is not possible to force all operators into a single permitting system, thus potentially undermining its benefits.
- Difficulties in undertaking detailed assessments of the relative impacts, etc., of each activity and, therefore, establishing integrated permit conditions.
- In complex situations, very complex permits might result, causing difficulties in interpretation for operators and regulators.
- Difficulties in identifying sources responsible for offences.
- Difficulties in implementing legal obligations.

## 13.6 CONSTRAINTS AND OPPORTUNITIES FOR INTEGRATED REGULATION OF ESTATES

There are limited opportunities for some aspects of estate-wide permitting where there is clear technical connection, but the number of such sites is limited. In some cases the building of a new industrial estate could be subject to an environmental impact assessment (e.g., in Austria), but while this would give some integrated view of activities and impacts at this stage, it would not lead to single integrated permitting for ongoing control of pollution nor for changes in the operations on the estate over time.

- A workshop established under the 'Exploring New Approaches' initiative on regulation (ENAP, 2004) concluded that the opportunities for introducing single permits were somewhat limited. In many cases it would be better to focus on voluntary co-operation and seeking links with other environmental management regulations. Overall, the opportunities that arise from taking an estate-wide approach to regulation are:
- Voluntary co-operation – where operators agree to co-operate in assessment and control, certain legal and regulatory hurdles can be overcome.
- Co-ordination of permitting processes for installations within an estate – such co-ordination can optimise controls, increase positive outcomes for the environment and reduce burdens and costs for business.
- Establishing requirements during the planning of an estate so that all new investors need to comply with these – this provides a level playing field and up-front certainty for operators. It also helps avoid 'free loaders' on a site.

- Linking various decision-making processes, including planning, building permits, accident management, IED, etc., helping to identify and deliver synergies.
- Opportunities might be most seen for new estates as integrated approaches across an estate can affect common design features. Further operators 'arriving' together may be more persuadable to work together than existing operators used to existing regulatory requirements.
- Synergies on regulatory approaches can deliver synergies for business opportunities. For example, linking controls on wastewater treatment can lead to common investments, saving cost and improving discharges.

The ENAP project identified a number of constraints that exist in one or more EU Member State in issuing permits covering all or much of an industrial estate:

- Legal constraints:
  - Current legislation at national level is entirely focused on specific installations and it is not possible to go beyond this.
  - The need for the permit holder to be a single legal entity, whereas an estate is usually only a geographic location with several legal entities.
- The problem of identifying, allocating and enforcing responsibility, if a single permit were issued covering more than one operator.
- Some owners would resist what might be viewed as an extension of regulation beyond what they might be otherwise legally subject to (e.g., to include activities that fall below thresholds required for inclusion in the IED). Such operators could argue that each activity on an estate should be regulated on its own merits and an 'estate' approach could run counter to the need for proportionate legislation.
- It might be difficult to ensure that each operator accepts 'ownership' of environmental issues if they are collectivised.
- It could lead to a potentially complex regime that could be confusing for operators and regulators.
- If there were to be a permit for the entire estate instead of single installations, this could make interpreting BAT in such a context especially challenging and could imply a move away from BAT and could lead to a lower level of environmental protection.
- Permitting itself is more difficult when broadening the scope but a real integrated way of handling the permit procedure can only be done this way.

## 13.7 OTHER INSTRUMENTS TO IMPROVE PERFORMANCE OF INDUSTRIAL ESTATES

A number of other instruments could be used to assist in the development of permits for industrial estates and/or treatment of environmental issues. Estate-wide Environmental Impact Assessments (EIAs) would require common assessments of issues by all of the operators and the regulator and could provide a basis for further regulation, whether this is a single permit, co-ordinated permits or a combination of measures to address pressures to meet WFD objectives. Furthermore, an operator, when carrying out an impact assessment of his installation, has to take into account the other installations and evaluate the cumulated impact on the environment and on the health of the surrounding population. The Seveso III Directive requirement for risk assessment, the use of domino effect evaluation and prevention, co-ordinating security management systems and emergency plans are required. At a national level the use of economic instruments (e.g., tax incentives) could be used to encourage units to work together as can other mechanisms to encourage a voluntary approach – which could involve not just negotiated agreements, but the development of estate-wide management systems. Areas of particular benefit could be managing common environmental services (e.g., wastewater treatment), encouraging efficient use of natural resources and waste (e.g., abstracted water), and addressing emissions that have a cumulative effect on local water bodies.

An estate-wide approach could, therefore, result in environmental benefits, improved capacity of institutions and public understanding. It would also stimulate a feeling of shared responsibility and progress towards new ways of achieving environmental outcomes. However, there is concern over the regulatory costs, especially whether it will be a disproportionate burden to the lower impact units within an estate.

It is also important to note that even where estate-wide permitting is not considered appropriate, other measures can be used to deliver some of the benefits that could be achieved by this approach. Simply processing permit applications at the same time can be beneficial. In assessing the impacts of emissions under the IED, for example, it is appropriate to consider the relative contribution of neighbouring sources and, therefore, determine where reductions in emissions are most appropriate (to be set out in individual permits). In Sweden this can be taken a stage further, where neighbouring installations can undertake to take joint measures to comply with environmental quality standards and, as a result, joint permit conditions are established.

## 13.8 CONCLUSIONS

An integrated approach to the control of pollution from industrial estates has many advantages in delivering the objectives of the WFD and providing a coherent and cost-effective approach to the operators of activities on an estate. However, there are constraints and opportunities to do this within EU law, including with regard to the IED.

Participants at the ENAP workshop (2004) agreed on 10 'Prague Principles', the first six of which are relevant to regulation of industrial estates and are set out in the box below (updated to refer to the IED).

**THE 'PRAGUE PRINCIPLES' RELEVANT TO REGULATION OF INDUSTRIAL ESTATES**

(1) The IED is the key piece of community legislation ensuring that the environmental aspects and impacts of industrial installations are properly regulated.
(2) The flexibility in the key terms of the IED, e.g., 'installation', 'operator' and 'permit', allows tailor made solutions in the different member states. This flexibility should be instrumental to protecting the environment. Exchange of best practices would be useful.
(3) Clarification of the way in which Member States interpret the terms 'directly associated activities', 'technical connection' and 'site' in the definition of 'installation' is desirable.
(4) A useful starting point for defining the 'scope' of the IED permit is one permit covering technically connected and/or directly associated activities on the same site under the control of the same operator. From there benefits of and preconditions for wider scoped permits can usefully be explored.
(5) Disintegration of installations and 'salami slicing' could lead to less effective regulation of industrial sites.
(6) To manage industrial estate issues, various regulatory and voluntary arrangements can be used to take account of any interactions between different activities.

In the context of the IED much of the debate concerning industrial estates concerns the flexibility available in the directives and the national legal context and, as a result, what can be brought together in a single regulatory process. In contrast, the WFD focuses on the recipient water body and requires an integrated assessment of the multiple pressures on that water body. The WFD requires that appropriate measures are adopted to address those pressures, but does not prescribe the nature of the regulatory processes to achieve this.

The assessment of pressures arising from an industrial estate, for example diffuse pollution, under the WFD requires an integrated analytical process if the control of such pressures is to be translated into measures which are effective, and proportionately and fairly distributed to the operators on that estate. With regard to the IED, there are issues of what can or cannot be included within an IED permit, the potential burden of 'over regulation' for small activities captured in a single permit, as well as sharing the responsibilities for meeting permit conditions. Having said this, the integrated thinking encouraged by the WFD to address pressures on water bodies and innovative thinking on the regulation of industrial estates can be brought together to find innovative and cost-effective solutions to the management of these sites.

## 13.9 ACKNOWLEDGEMENTS

This chapter draws on work undertaken by the author for the EU Network for the Implementation and Enforcement of Environmental Law (IMPEL). The author is grateful to IMPEL for this support and for interesting discussions with IMPEL members from many Member States. It should be noted, however, that the views expressed here are those of the author alone.

## REFERENCES

EC (2007). Guidance on Interpretation of "Installation" and "Operator" for the Purposes of the IPPC Directive. European Commission, Brussels.

ENAP (2004). Workshop 3. Exploring the Scope of Permits Under the IPPC Directive and Alternative Approaches for Regulating Industrial Activities – Opportunities and Constraints. Workshop Conclusions. VROM, The Netherlands.

EU (2000). Directive 2000/60/EC of the European Parliament and of the Council establishing a framework for Community action in the field of water policy. OJ L327, December 22, 2000.

EU (2010). Directive 2010/75/EU of the European Parliament and of the Council on industrial emissions (integrated pollution prevention and control). OJ L 334, December 17, 2000.

Farmer A. M. and Cherrier V. (2010). Linking the Water Framework Directive and the IPPC Directive. Phase 1. IMPEL, Brussels.

Farmer A. M. and Cherrier V. (2011). Linking the Water Framework Directive and the IPPC Directive. Phase 2. IMPEL, Brussels.

# Chapter 14

# The regulatory regime for bringing SUDS into routine use for industrial estates and business parks in Scotland, UK

*Brian J. D'Arcy[1]* and Andrew Hemingway[2]*

*[1]Independent Environmental Consultant & Research Fellow Abertay University, UK*

*[2]SEPA*

**Corresponding author: info@enviroexperience.co.uk*

## 14.1 INTRODUCTION TO THE PROBLEM

The environmental problems of stormwater pollution from industrial estates have been described elsewhere (e.g., D'Arcy & Bayes, 1995; Scottish Environment Protection Agency [SEPA], 1999; D'Arcy *et al.* 2000). This chapter is focused on how to bring about appropriate control measures and practices. The latter are known as BMPs, or Best Management Practices in the USA and many other countries, and comprise behaviours and physical structures. In the UK, the use of the physical measures (often referred to as urban BMPs) became known as sustainable drainage (D'Arcy, 1998) and hence as Sustainable Urban Drainage Systems, or SUDS, and is seen as integral to management of stormwater (D'Arcy, 2013). How to bring them into routine use (including control of pollution risks to allow them to be effective) is the subject of this chapter.

Development of a regulatory regime for stormwater management at commercial and industrial premises is complex. The wide variety of pollutants to be controlled varies with the industries present in any particular area, and varies with time. Whilst there is scope for restrictions on the use of some pollutants (e.g., timber treatment chemicals, toxic metals in automobile components, polycyclic aromatic

© IWA Publishing 2017. Wealth Creation without Pollution: Designing for Industry, Ecobusiness Parks and Industrial Estates
Brian D'Arcy, Lee-Hyung Kim, Marla Maniquiz-Redillas
doi: 10.2166/9781780408330_205

hydrocarbons [PAHs] in sealants), there will always remain a need for additional measures to manage pollution risks from every day contaminants such as hydrocarbons, nutrients and suspended matter including soil. Polluting activities are very variable. These range from vehicle emissions, vents opening onto roof areas depositing contaminated dust from factory units, and seepages from casual storage of wastes and raw materials, to spillages and larger scale accidents, as well as illicit dumping of pollutants into drainage systems. Rainfall mobilises all those pollutants into and from drainage networks, giving the problem a diffuse pollution character (Novotny & Olem, 1993; D'Arcy *et al.* 2000; Ferrier *et al.* 2005).

The ability of a regulatory regime to prevent pollution from those substances and activities is dependent on promotion of best practices, including drainage infrastructure (SUDS) that can capture diffuse sources of contamination. But how to integrate stormwater flooding issues with regulatory needs for water quality? Which organisations should have responsibility for which issues? If landscape features have drainage functions, who does the maintenance at an estates scale where shared drainage structures are involved, and from whose budget? How to prevent overloading of the treatment systems? In addition to the kind of diffuse pollution sources prevalent for other built environments (road runoff, casual disposal of waste such as engine oil, paint thinners, miscellaneous wash waters, etc.), industrial catchments include sources which can involve trade effluent misconnections, careless or inadequate storage of wastes, and other activities, all of which can come under a range of statutory controls.

Analysis of opposition to SUDS technology in the UK identified multiple benefits as one means of overcoming barriers (D'Arcy & Frost, 2001). The aspirations for the technology therefore include flood risk management and also broader interests loosely included as amenity, but encompassing landscape appearance, biodiversity, and contributions to well-being in towns and cities (D'Arcy, 2013). Multiple benefits are sought after from public sector expenditure, but multiple funding is not usual. Figure 14.1 indicates the complexity of the regulatory challenge for stormwater runoff.

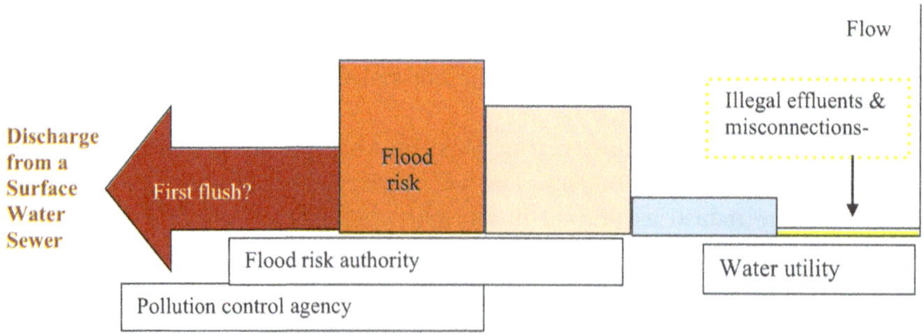


**Figure 14.1** The regulatory hydrograph – a discharge from a surface water sewer draining an industrial estate indicating the mixture of different problems, with consequent different remits for various agencies.

## 14.2 THE BASIS OF THE REGULATORY APPROACH

The regulatory regime for an environmental problem needs an analysis of the nature of the problem: what are the principal impacts and causes, which businesses and sectors should be held responsible for which aspects? And who should be a licensing or other control body?

Details of the infrastructure of industrial estates obviously have implications for the development of a regulatory regime for the stormwater runoff. Figure 14.2 shows how a modern industrial estate is usually drained on a separate sewer system, with the stormwater draining to watercourse, and the foul drainage collected separately and taken to a treatment works.

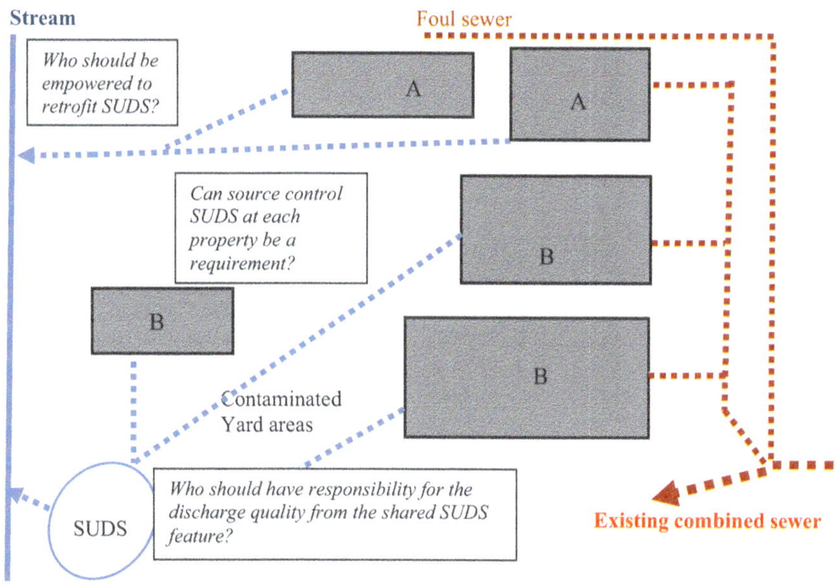


**Figure 14.2** The regulatory challenge for an example industrial estate with existing infrastructure (factory units A) with no SUDS, and new development (units B) including SUDS features. (Source control SUDS features not shown.)

Any regulatory regime needs to be fair and effective, with sensible enforcement options. (Campbell *et al.* 2004). For immediate regulatory action, existing legislation usually offers some scope for action, and experience using existing and earlier legislation can also inform new measures. In Scotland there are some privately developed estates, but usually drainage is into public sewers that are the responsibility (including discharge quality) of the water utility. But pollution risks are from a variety of sources, including major incidents at particular premises over which the water utility has little control.

Figure 14.3 considers how the pollution risks are spread across an industrial estate, and how the pollutants get from source to watercourse, together with scope for SUDS interventions. Some of the roles of various regulatory agencies for control and prevention of problems are indicated; the basis for development of an appropriate regulatory regime. The environment agency has a remit for each category in Figure 14.3, and is a statutory consultee for local authority planning consultations on new developments: a key player.

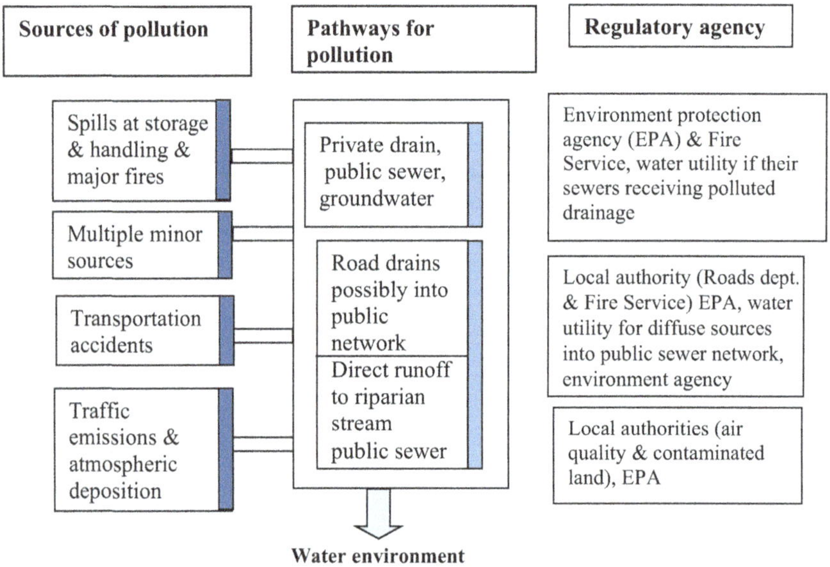


**Figure 14.3** Analysis of the regulatory challenges for industrial estates: infrastructure interventions can be at source, within the pathways, and immediately prior to discharge into the water environment; 3 'levels of treatment'.

Every treatment system has a finite capacity to accept influent pollution loads, and exceeding the design influent loading will cause failures. SUDS are not designed to receive high levels of contamination. Therefore, it was recognised in developing a regulatory regime for industrial estates (and indeed for other urban diffuse sources of pollution) that if such a regime is to require pollutant capture and attenuation features as part of the drainage infrastructure, and those features are to have amenity and even wildlife value, then the regulatory regime needs to prevent disposal of pollutants into the surface water drainage system itself. That goes beyond simple oil storage regulations, and seeks to protect surface water drainage networks from sewage and trade effluent misconnections or pumping station failures, as well as chemical or other pollutant leaks and spills.

Two different regulatory decisions need to be taken if use of SUDS technology is to become established as routine, and stormwater environmental problems are to gradually diminish:

(a) To stop the environmental degradation problem from continuing to grow with each and every new development, a regulatory regime is needed which requires new developments and redevelopments to use SUDs technology and thereby address diffuse pollution.
(b) For effective maintenance of shared SUDS, a public body needs to be given responsibility for SUDS in the public domain. And in order to make inroads into the existing environmental problems, that public body must have the remit to spend public money on retrofit programmes.

Those two separate regulatory actions are illustrated in Figure 14.4.

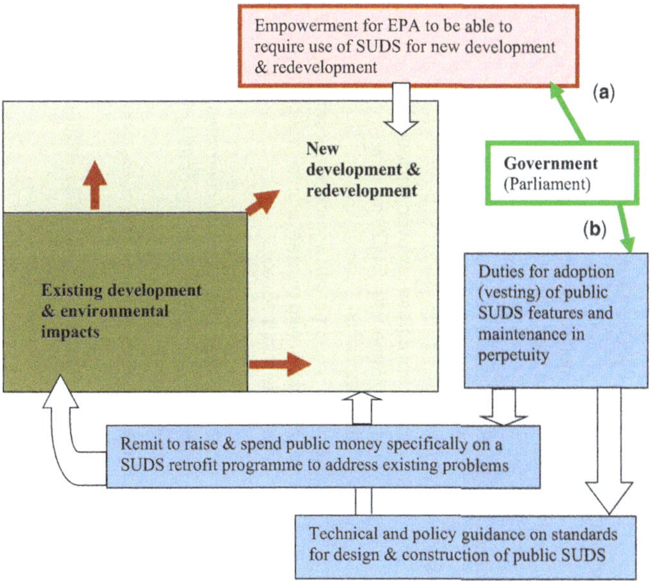


**Figure 14.4** Dual focus of a regulatory regime to bring SUDS technology into routine business: (a) empowerment for an environment protection agency (EPA) to require use of the technology, and (b) remit for a public body (e.g., water utility) to adopt and maintain it as public infrastructure.

## 14.3 RESULTS

### 14.3.1 Legislation in Scotland and SUDS technology

The enabling legislation is summarised in Table 14.1, starting with the Control of Pollution Act 1974, (COPA), which preceded the development of diffuse pollution

**Table 14.1** Enabling legislation in Scotland for controlling diffuse pollution from the built environment, including establishing a requirement for SUDS technology.

| Legislation | Target Impact for SUDS | Regulatory Scope & Remits |
|---|---|---|
| Control of Pollution Act, 1974<br>Provisions used from 1994, in Forth catchment, and from 1996 across Scotland | Allowed discharges of surface water to be controlled by a formal consent (permit) | To require treatment for surface water from new developments. Used by Forth River Purification Board then SEPA. |
| Planning legislation:<br>(1) development control<br>(2) strategic plans | To deliver multiple benefits; amenity, pollution control and flood risk management. | Required developers to use innovative technology to manage stormwater for environmental protection and enhancement |
| Water Environment and Water Services (Scotland) Act 2003 (the WEWS Act – implementation of the Water Framework Directive in Scotland). | (1) Resolve confusion about adoption and long term maintenance of SUDS features<br>(2) Provide for regulations to be written to control diffuse pollution, including a requirement for SUDS for new developments | (1) Scottish Water remit for public SUDS with powers to write vesting requirements; establish a clear remit for water utility role.<br>(2) Establish a much simpler regulatory basis for SEPA to require use of SUDS, and to protect surface water and the water environment |
| The Water Environment (Controlled Activities) (Scotland) Regulations 2005 (CAR) | Provide a proportionate approach to controlling pollution risks, with minimal administration for individually minor sources | Established a regulatory regime with 3 tiers:<br>Licence<br>Registration<br>General Binding Rules (GBR) |
| CAR GBR 10. Discharge of surface runoff from a surface water drainage system to the water environment | Control drainage from construction sites, buildings, roads, yards and any other built developments | Establish pollution prevention requirements including use of SUDS for new developments |
| CAR GBR 11. Discharge into a surface water drainage system | Protect amenity value of SUDS, especially habitat | Allow pollution control at source, not just in river |
| Nature Conservation Act 2004 | Promotion of the biodiversity possibilities of SUDS | Water utility, local councils and SEPA should promote biodiversity in policy & action |
| Flood Risk Management (Scotland) Act 2009 | Requires recording of stormwater attenuation assets by local councils | Use SUDS for pluvial floods (remit for local authorities) |
| Building Control regulations 2005 | Plot by plot measures to drain premises; structural integrity (including pollution prevention) | SUDS proposals on individual plots can be checked by local council at construction stage |

concepts and appropriate measures to address that problem, but did establish statutory means to control surface water discharges. The Environment Agency (EA), Scottish Environment Protection Agency (SEPA) and their predecessors often licensed industrial estate surface water dischargers, where individually significant estates impacted watercourses. For example, in 2011 there were 655 Industrial Estates with Discharge Consents in England and Wales; 617 in England and 38 in Wales (Environment Agency, 2011). The powers available under COPA and preceding legislation however, did not easily allow control at source; at individual premises.

## 14.3.2 Implementation of the Water Framework Directive in Scotland

The EU Water Framework Directive, WFD, was implemented in Scotland by the WEWS Act (2003). As outlined above, that legislation established two important new positions in law in Scotland:

(1) It gave Scottish Water the remit for public SUDS, and empowered the water utility to produce design-and-construct standards that would have to be met if a system was to be vested with Scottish Water.
(2) It established the requirements for pollution prevention for diffuse sources (via GBRs and licences), including a simple requirement for new developments to use SUDS technology.

### *14.3.2.1 The responsibility for shared ('public') SUDS features*

The need for SUDS technology had been driven by regulatory activity under the somewhat convoluted provisions of COPA since 1996 across Scotland. But difficulties concerned with ownership and maintenance responsibilities were unresolved, and by the time the WEWS Act was being drafted, action to clarify remits was needed. The national stakeholder working party for implementation of the technology debated the ideal scenario, and voted to have a remit established under the provisions of the WEWS Act 2003 for Scottish Water. That had the additional merit of giving a stronger remit to the water utility for retrofits, including establishing a budget in their capital programme.

The new statutory position also empowered Scottish Water to write regulations setting out details for adoption (vesting) of SUDS – conditions that would determine whether a feature was fit for purpose prior to being vested in Scottish Water. The hope was that such technical guidance would drive up standards, and that the chances of achieving fit for purpose SUDS would greatly improve. The standards were included in a re-written version of the earlier water industry guidance policy document on Standards for Sewers.

#### 14.3.2.2 A requirement for new development to use SUDS techniques

The outcome of the question 'How to require SUDS?' was determined by the Scottish Government in close consultation with SEPA and other stakeholders. The regulatory framework that was developed for authorising any drainage discharge under the WEWS Act comprised three options:

(a) Discharge licence
(b) Registration
(c) General binding rules

The discharge *licence* allowed for detailed site specific conditions including SUDS design detail and potentially numerical standards for a variety of pollutants. It was established for the most significant and highest risk discharges. The surface water drainage from industrial estates (where collected to form a single major discharge) was accepted as appropriate for a licence under that category. Licensing enables the environmental regulator to undertake monitoring and therefore collect evidence to prioritise SUDS retrofits by the licence holder for the drainage system (Scottish Water, as above). In that way a mechanism for improving the baseline, unavoidable levels of contamination that were part of the existing chronic pollution problem, was established.

*Registration* was established to provide a record of lower risk discharges such as small sewage treatment installations, septic tanks etc. and requires dischargers to register descriptive details of their discharge and property, including a map reference.

For most surface water discharges the third option, *general binding rules*, was adopted. The general binding rules (GBRs) concept was developed to bring diffuse sources under some kind of control. The vastly larger number of such sources precludes either of the other options; it would clearly be impossible to regulate every individual surface water drain by written permits or licences and associated individual meetings and sampling. The administration effort to register every such discharge would also be disproportionate to the environmental risk presented by each one. Instead simple descriptive requirements were established to protect the water environment from diffuse sources and to specify a legally binding requirement for new developments to use SUDS technology. Those specific requirements (see Discussion below) could be enforced by serving notices to require action, with enforcement action through the courts if needed. The GBRs and licensing regime in Scotland were established by The Water Environment (Controlled Activities) (Scotland) Regulations 2005, referred to henceforth as CAR (see description in D'Arcy *et al.* 2006).

The following activities are controlled by General Binding Rules (GBR) when not thought to be individually significant:

- Discharge of surface runoff from a surface water drainage system to the water environment from construction sites, buildings, roads, yards and any other built developments (GBR 10).
- Discharge into a surface water drainage system (GBR 11).

GBR numbers 10 and 11 are primarily to prevent pollution, or damage to SUDS treatment systems. The pollution prevention requirements in GBRs 10–11 refer to surface water drainage from the built environment and may be described loosely as follows:

- The discharge shall not contain any trade effluent or sewage, shall not cause visual impairment, destabilisation of stream bed or banks, or cause pollution.
- SUDS technology is to be used for new developments.
- The discharge shall not contain drainage from new areas where chemical, oil or other polluting matter is (un)loaded or handled.
- Pollutants shall not be disposed of into a surface drainage system. No material that might impair its performance and function shall be allowed to enter a surface water drainage system (includes SUDS).

As an example of the approach (minimal prescriptiveness; simple enforceable requirements), details of the GBRs in force in Scotland in 2015, are given in Appendix. Regulatory guidance for SEPA has been produced (WAT-RM-08).

### 14.3.3 Design and construction

The effectiveness of SUDS technology can be compromised if insufficient attention is paid to design and construction. In Scotland a building warrant is required for the completion of a development approved under planning legislation. The Building Control function of the local authority is concerned with structural and functional integrity of buildings and property. Revisions to Building Control regulations in 2005 added provisions to align the inspection and checking remit with the pollution and stormwater management requirements under consideration at that time by Scottish Government (e.g., the WEWS Act and subsequent regulations).

### 14.3.4 Managing flood risk

In 2007, severe flooding in the UK resulted in renewed focus on the role of the local authorities to take actions to manage flood risk more effectively than hitherto, especially given the likely impacts of climate change. The Flood Risk Management (Scotland) Act 2009, helps integrate flood risk aspects of stormwater with pollution prevention. The local authorities now have statutory requirements to prepare plans to manage flood risk, providing the framework for co-ordinating actions across catchments. Included in the obligations under the Flood Risk Management Act, is a requirement for each local authority to record flood risk assets (including SUDS features). SUDS are now recognised as essential elements in managing pluvial flood risks (i.e., due to rainfall).

## 14.4 DISCUSSION

### 14.4.1 Enforcement

Regulations without enforcement are self defeating, creating in the regulated sector a disregard for any requirements. A common monitoring strategy for major point sources involves site visits for all individual premises above a certain size, and reducing the frequency of visits for smaller sites on a pro-rata basis. This approach means individual surface water discharges and SUDS would rarely or never be inspected. Yet the collective impact of surface water drainage exceeds the impact of the regularly sampled and inspected effluent discharges (SEPA, 1999). For the many surface water discharges authorised by GBRs, an effective enforcement policy may require a strategy involving inspection of a small *proportion* of discharges, varying the locus each year. This planned programme of inspections could be funded by the direct funding for the agency. The existence of such a programme (but not the locations) should be publicised as should all enforcement actions. A degree of cost recovery for enforcement could be achieved by use of administrative penalties, such as those proposed in Scotland in 2016 (and see D'Arcy *et al.* 2006).

### 14.4.2 The relationship between statutes and policy

It is not a simple and speedy option to bring in new legislation. Whenever so doing therefore, a balance is needed between:

(i) establishing clear unequivocal (and hence enforceable) statutory requirements, and
(ii) using regulation to establish a framework for actions, details of which can be left to government policy to guide interpretation.

A second level of policy arises when the regulatory agencies write their own informal policies to interpret government policy for the guidance of their staff and the regulated sectors. Such guidance should never, of course, undermine the will of the legislature in establishing legal requirements. Problems can arise when government establishes duties and gives powers, but internal policy priorities and culture in the responsible bodies work against those statutory duties.

There is considerable scope for development of more effective flow and pollutant attenuation techniques for stormwater management. But how to bring innovative technology into routine use without unnecessary risks for the environment, or conversely stifling much needed inventiveness? Design details are not beyond the reach of regulation, but care is required to avoid regulatory requirements that are over-prescriptive; requirements beyond a basic requirement for the technology are generally best delivered by working with sectors to ensure detailed guidance and policies are going to be effective.

### 14.4.3 How to achieve multiple benefits from SUDS technology?

SUDS has the potential to deliver amenity and biodiversity benefits as well as flood risk and water quality benefits. A holistic, co-ordinated approach to maximise benefits in each of these areas will deliver much better value for money. A major problem however for implementation of technology once it is brought in by a specific statute, is the 'tunnel vision' administration of the technology via that statute and that one alone. Stormwater is a single entity, but the implementation of two distinct EU Directives (e.g., Water Framework, and Floods) involves separate teams of administrators each with their own priorities. In addition, an over-arching statute, such as the duty imposed on public bodies in Scotland to promote biodiversity in the course of all their duties by the Nature Conservation Act 2004, is not owned and championed by anyone in those functional teams.

### 14.4.4 Sector engagement and economic drivers

The light touch regulatory approach exemplified by non-registration GBRs, as described above, means that the pollution prevention requirements have to be well publicised to the regulated sectors. That is the hidden cost of light touch regulation. There remains a need for serious awareness-raising campaigns if a new regulatory regime is to be effective. By comparison with many cities in the USA (see USEPA, 2008), the effort put into public education to support the regulatory achievements in the UK is too limited. The programmes of measures that have been developed in catchments for implementation of the Water Framework Directive in the EU provides a focus for awareness-raising in Scotland and support for the regulatory measures. Economic drivers will also help and have been implemented, for example in Berlin, in Germany, whereby a household may save 20% on water charges by disconnection of surface runoff. Similar schemes have been introduced in Seattle and other cities in USA. For non-domestic customers such as industrial estate occupants, the water utility in Scotland, Scottish Water, is considering such economic instruments. That would complete the regulatory framework set out in Figure 14.3 above. An often-overlooked economic consequence of regulatory requirements is the creation of a stable market for expertise and for new products (unless undermined by failures of statutory bodies to apply the law).

## 14.5 CONCLUSIONS

The difficulties to be overcome in bringing into existence an effective regime for regulating stormwater, in particular making it a legal requirement to use sustainable urban drainage systems (SUDS), include the diffuse source nature of pollution risks, the great variety of pollutants, and the challenge to integrate pollution prevention with flood risk management and broader aspirations for landscape and amenity. An effective regime needs to target pollution sources, but be applied in an even handed and fair way. As well as direct regulation, economic incentives and stakeholder engagement techniques are important in bringing the technology into use and achieving fit for purpose features. The technology can be compromised if insufficient attention is paid to design and construction; details which are deliverable by working with sectors to ensure guidance and policies. Effectiveness in the longer term requires parallel regulatory efforts to address persistent and toxic pollutants at source. Any regulatory regime requires an enforcement strategy if it is to be effective. The variety of organisations with partial responsibility for aspects of the stormwater management challenge makes development of appropriate enforcement regimes a little more difficult than traditional regulatory activities, and requires a degree of innovation in approach.

## 14.6 ACKNOWLEDGEMENTS

The help of Ian Speirs, Gaye McKay, Bridget Woods Ballard and Lee Hyung Kim in commenting on drafts is gratefully acknowledged.

## REFERENCES

Bayes C. D. and D'Arcy B. J. (1995). Industrial estates – how to prevent pollution. In: C. Pratt (ed.), Proceedings of the Tenth Meeting of the Standing Conference on Stormwater Source Control. School of the Built Environment, Coventry University, Coventry, ISBN 0 905949 33 1.

Campbell N. S., D'Arcy B. J., Frost C. A., Novotny V. and Sansom A. L. (2004). Diffuse Pollution – an Introduction to the Problems and Solutions. IWA Publishing, London, ISBN: 1 900222 53 1.

D'Arcy B. and Frost A. (2001). The role of best management practices in alleviating water quality problems associated with diffuse pollution. *The Science of the Total Environment*, **265**, 359–367. Elsevier

D'Arcy B. J. (1998). A New Scottish Approach to Urban Drainage in the Developments at Dunfermline. Proceedings of the Standing Conference on Stormwater Source Control. Vol. **XV**. The School of the Built Environment, Coventry University, Coventry.

D'Arcy B. J. (2013). Managing Stormwater: from aspirations to routine business. International Issue 2013. *gwf-Wasser/Abwasser*, **154**, gwf, S1/2013, ISSN 0016-3651, B 5399.

D'Arcy B. J., Ellis J. B., Ferrier R. C., Jenkins A. and Dils R. (2000). Diffuse Pollution Impacts: the Environmental and Economic Impacts of Diffuse Pollution. Chartered Institution of Water and Environmental Management, London, ISBN: 1 870752 46 5.

D'Arcy B. J., Ellis J. B., Ferrier R., Jenkins A. and Dils R. (2000). Diffuse Pollution Impacts: the Environmental and Economic Impacts of Diffuse Pollution in the UK. CIWEM, London. ISBN: 1 870752 46 5.

D'Arcy B. J., Schmulian K. and Wade R. (2006). Regulatory options for the management of rural diffuse pollution. In: Managing Rural Diffuse Pollution, L. Gairns, K. Crighton and B. Jeffrey (eds), Proceedings of the SAC & SEPA Biennial Conference, 5–6 April 2006, Edinburgh, pp. 192–200, ISBN 1-901322-63-7.

Environment Agency (2011). Freedom of Information Act enquiry, ref. no. NRJUL53, reported Monday, 1 August 2011.

Ferrier R., D'Arcy B. J., MacDonald J. and Aitken M. (2005). Diffuse pollution – what is the nature of the problem? *Journal of the Chartered Institution of Water and Environmental Management WEJ*, **I**, 361–366.

Novotny and Olem (1994). Water Quality: Prevention, Identification and Management of Diffuse Pollution. Van Nostrand Reinhold, New York, reprinted and distributed by J. Wiley and Sons, New York.

SEPA (1999). Improving Scotland's Water Environment – SEPA State of the Environment Report. Scottish Environment Protection Agency, Stirling.

SEPA (2015). Regulatory Method (WAT-RM-08) – Sustainable Urban Drainage Systems (SUDS or SUD Systems). Scottish Environment Protection Agency, Stirling; USEPA (2008) Handbook for Developing Watershed Plans to Restore and Protect Our Waters. EPA 841-B-08-002, March 2008, Washington DC. USEPA web site: http://iaspub.epa.gov/waters10/attains_nation_cy.control

## APPENDIX

General binding rules from controlled activities regulations, Scotland.

| Activity | Rules |
|---|---|
| (10) Discharge of water run-off from a surface water drainage system to the water environment from buildings, roads, yards or any other built developments, or construction sites for such developments, and, if desired, the construction and maintenance of any water outfall in or near to inland surface water which forms, or will form, part of that system. | (a) All reasonable steps must be taken to ensure that the discharge must not result in pollution of the water environment;<br>(b) the discharge must not contain any trade effluent or sewage, and must not result in visible discolouration, iridescence, foaming or growth of sewage fungus in the water environment;<br>(c) the discharge must not result in the destabilisation of the banks or bed of the receiving surface water;<br>(d) the discharge must not contain any water run-off from any built developments, the construction of which is completed after 1st April 2007, or from construction sites operated after 1st April 2007, unless—<br>(i) during construction those developments are drained by a SUD system or equivalent systems equipped to avoid pollution of the water environment;<br>(ii) following construction those developments are drained by a SUD system equipped to avoid pollution of the water environment;<br>(iii) the run-off is from a development that is a single dwelling and its curtilage; or<br>(iv) the discharge is to coastal water;<br>(e) the discharge must not contain any water run-off from<br>(i) fuel delivery areas and areas where vehicles, plant and equipment are refuelled; |

*(Continued)*

General binding rules from controlled activities regulations, Scotland (*Continued*).

| Activity | Rules |
|---|---|
| | (ii) vehicle loading or unloading bays where potentially polluting matter is handled; or<br>(iii) oil and chemical storage, handling and delivery areas; constructed after 1st April 2007;<br>(f) all facilities with which the surface water drainage system is equipped to avoid pollution, including oil interceptors, silt traps and SUD system attenuation, settlement and treatment facilities, must be maintained in a good state of repair;<br>(g) all reasonable steps must be taken to ensure that any matter liable to block, obstruct, or otherwise impair the ability of the surface water drainage system to avoid pollution of the water environment is prevented from entering the drainage system; and<br>(h) the construction or maintenance of the outfall must not result in pollution of the water environment. |
| (11) Discharge into a surface water drainage system. | (a) Oil, paint, paint thinners, pesticides, detergents, disinfectants or other pollutants must not be disposed of into a surface water drainage system or onto any surface that drains into a surface water drainage system;<br>(b) any matter liable to block, obstruct, or otherwise impair the ability of the surface water drainage system to avoid pollution of the water environment must not be disposed of into a surface water drainage system or onto a surface that drains into a surface water drainage system;<br>(c) sewage or trade effluent must not be discharged into any surface water drainage system; and<br>(d) on construction sites any area of exposed soil from which water drains into a surface water drainage system, and the period of time during which such water drains, must be the minimum reasonably necessary to facilitate the construction works being undertaken at that site. |

# Chapter 15

# Regulatory regimes for diffuse pollution and industrial estates in Korea

*Lee-Hyung Kim*

*Department of Civil and Environmental Engineering, Kongju National University, Cheonan, Chungnamdo, Korea (E-mail: leehyung@kongju.ac.kr)*

## 15.1 INTRODUCTION

Pollutants from point and nonpoint sources discharge in watershed areas contribute to the deterioration of surface water quality and the aqua-ecosystem. Pollutants generated from point sources such as wastewater treatment plants in urban, industrial, and livestock areas have constant quantity and quality during both dry and wet seasons; contrary to nonpoint sources (NPSs). Wastewaters from point sources are site-specific and can be easily collected and treated especially when the wastewater collection system has already been well established. The Republic of Korea has made a tremendous advancement in the past 40 years to clean up the aquatic environment by controlling pollution from point sources (Kim, 2007). More than 90% of the country's wastewater is being collected in sewerage systems and treated in wastewater treatment plants. But even though point source discharges have decreased during the recent decades, the aquatic-ecosystems are still contaminated with excess algal biomass, turbid waters, toxic chemicals, etc. due to NPS pollution. There are many potential sources of NPS pollution including agriculture, forestry, grazing, septic systems, recreational boating, urban and industrial stormwater runoff, construction, physical changes to stream channels, and habitat degradation. NPS pollution in urban, agricultural, industrial, livestock, mining areas, etc. are more difficult to detect and manage since pollutants could be

© IWA Publishing 2017. Wealth Creation without Pollution: Designing for Industry, Ecobusiness Parks and Industrial Estates
Brian D'Arcy, Lee-Hyung Kim, Marla Maniquiz-Redillas
doi: 10.2166/9781780408330_219

widely discharged from any type of land use during storm events. The stormwater runoff carries away natural and anthropogenic pollutants. The NPS pollutants are deposited into lakes, rivers, wetlands, coastal waters, etc. (Kim, 2003). Both land use and climate change can contribute to NPS pollutant load to the rivers by increasing the runoff and rainfall intensity, and decreasing infiltration to the soil. The change of the natural hydrologic cycle in the watershed areas can also influence the types of discharged pollutant. High rainfall intensity and peak flows can mobilize large particles with high energy potential.

The chemical oxygen demand (COD), which is a parameter indicating the biodegradability and non-biodegradability of organic compounds shows an increasing trend even if the biochemical oxygen demand (BOD) concentration decreases (Figure 15.1). The discharge of non-biodegradable organic compounds is mainly due to NPS pollution from industrial, agricultural, and urban areas. Thus, NPS pollution remains a major cause of degradation of receiving waters in Korea. The Ministry of Environment (MOE) reported that the BOD mass loadings in the four major rivers in Korea ranged between 50% and 60% of the total BOD load in 2015. The MOE predicts that the BOD loadings could reach up to 70% by 2020 (MOE, 2013).

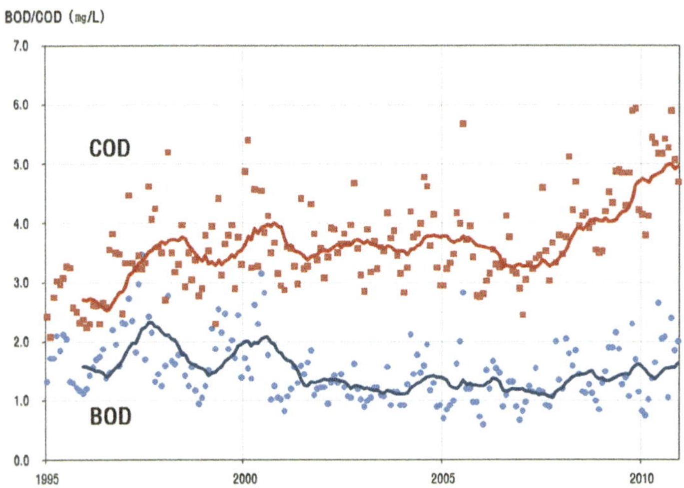


**Figure 15.1** Contribution of NPS pollutant loading rates in four major rivers in Korea (MOE, 2006).

Urbanization and industrialization significantly alter the natural hydrologic cycle by increasing the imperviousness rate. Imperviousness could result in increased runoff and peak flow, longer peak flow duration, high water temperature, and increased NPS pollutant discharges (Table 15.1). These hydrological and environmental changes could cause flooding, habitat damage, soil erosion, river channel widening, and streambed alteration; as well as deterioration of the surface water quality and destruction of the aqua-ecosystem. Thus, these changes are necessitating a robust regulatory regime for the effective management of industrial stormwater runoff.

**Table 15.1** Impacts from increase in impervious surfaces.

| Increased Imperviousness *Leads to:* | Resulting Impacts | | | | |
|---|---|---|---|---|---|
| | **Flooding** | **Habitat Loss** | **Erosion** | **Channel Widening** | **Streambed Alteration** |
| Increased Volume | • | • | • | • | • |
| Increased Peak Flow | • | • | • | • | • |
| Increased Peak Flow Duration | • | • | • | • | • |
| Increased Stream Temperature | | • | | | |
| Decreased Base Flow | | • | | | |
| Changes in Sediment Loadings | • | • | • | • | • |

*Source:* United States Environmental Protection Agency [US EPA], 1999.

In 2000, a significant change in the regulatory regime of water quality management in Korea took place with the implementation of the 'Total Pollutant Load Management System' (TPLMS), adopted from the concept of the 'Total Maximum Daily Load' (TMDL) program in the United States. The TPLMS is a program intended for the management of the total daily pollutant load input (point and nonpoint sources) from the watershed area to the rivers and lakes (MOE, 2016). After the introduction of TPLMS, the necessity for NPS pollution management greatly increased which resulted in the development and improvement of the legislations, regulations and management techniques concerning NPS. In this chapter, the regulatory regimes and management measures are discussed focusing on stormwater management in industrial areas.

## 15.2 COMPREHENSIVE MEASURES FOR NPS POLLUTION MANAGEMENT

Industrial areas are sites subjected to high usage of various chemicals in the manufacturing processes of certain goods. Such chemicals can contribute harmful effects to humans and the ecosystem with acute toxicity, fish toxicity, mutagenicity, oncogenicity, and others. For instance, chemicals such as dioxins, phenols and heavy metals can accumulate in the biomass but are not degraded in the natural environment. The long-term bioaccumulation can have fatal impacts on human health by damaging the immune system and interfering with hormones, impairing mental and neurological functions, and altering numerous metabolic body processes.

However, regulatory approaches to manage stormwater runoff from industrial areas are institutionally challenging, since it is not clear who owns the stormwater. The final discharge into receiving water contains emissions from all properties, and there is no clear holder of the responsibility to manage the stormwater runoff. Thus, regulations become more difficult to implement due to the complexity of the pollution source (Kim, 2003).

The 'Comprehensive Measures for NPS Pollution Management' was established in March 2004 by the MOE for the protection of the four major rivers in Korea. It primarily consists of regulation, policy, and development of technology and management in the six main fields including urban, agricultural, and forest areas, research, advertising and program. It was revised in 2012 with the extension of the third phase until 2020 considering the changes of water-environment, weather, economics and technologies (Table 15.2). The first phase is the period of building the policy foundation by pursuing pilot projects and identifying the causes of NPS pollution and technologies for NPS management. The second phase is the period of changing the management responsibility of pollution sources by providing best management practice (BMP) projects. The third phase is the period of continuing the strong management responsibilities by applying low impact development (LID) and green infrastructure (GI) techniques.

## 15.3 REGULATORY REGIME FOR DIFFUSE POLLUTION MANAGEMENT IN INDUSTRIAL AREAS

An industrial complex generates various pollutants from point sources and NPS during the production process. In addition, hazardous and toxic chemicals can be discharged from chemical accidents, spills, fire, etc. Highly impervious surfaces such as roads and parking lots in industrial sites also contribute to the increase of NPS pollutants that can accumulate during dry days on the paved surfaces and be washed-off by runoff during rainfall events. Thus, management of NPS pollution has become a vital environment policy in Korea.

### 15.3.1 Fundamental act of water cycle

Water provides numerous benefits to human life, industrial activity, and cultural development. However, the water cycle is altered because of urbanization, industrialization and climate change. Changes in the natural water cycle can lead to flooding, water pollution and ecological damage. Thus, developed countries, including Korea, have developed various policies and regulatory regimes considering water in recognition that water is a common property of all humans. The 'Fundamental Act of Water Cycle' defines water as '*an important share of the national wealth with high water publicity.*' It also specifies that integrated water management is necessary on a watershed scale in order to maintain a healthy water circulation and recovery.

**Table 15.2** Major goals of NPS pollution measures for each phase.

| Phase (Period) | Field | | |
|---|---|---|---|
| | **Policy System Improvements** | **Related Projects** | **Investigation and Research** |
| 1st Phase (2004 to 2005) | – Establish legislative requirements for NPS | – Perform national pilot projects to provide management directions | – Identify NPS pollution sources and develop management strategies |
| 2nd Phase (2006 to 2011) | – Apply NPS management for environment impact assessment (EIA) projects<br>– Establish reporting system of NPS management facilities | – Perform national best management practice (BMP) projects for site specific conditions in four major rivers to provide technological management approaches for pollutant removal and performance evaluation | – Develop treatment technologies for NPS pollution management<br>– Develop BMP design manuals |
| 3rd Phase (2012 to 2020) | – Establish fundamental act of water cycle<br>– Establish NPS control area designation system<br>– Implement strong regulatory regimes for NPS<br>– Require low impact development (LID) and green infrastructure (GI) technology for development projects<br>– Enact retention basin program for industrial areas to control stormwater runoff | – Perform water circulation city development projects, LID and GI development projects, NPS civic participation projects, and ecological river development projects | – Improve, verify and implement cost-effective BMPs<br>– Develop LID and GI guidelines<br>– Launch National NPS research team to develop LID/GI technologies |

*Source:* MOE (2012).

## 15.3.2 Total pollutant load management system (TPLMS)

In Korea, the population and industrial areas are excessively clustered in the middle and downstream section of the rivers. Following the introduction of the TPLMS in Article 4 of the 'Act of Water Quality and Aqua-ecosystem Protection,' it has been earnestly applied since 2004. In the TPLMS, the local government regulates the total pollutant emission load to achieve the target water quality standards in waterbodies. Development is restricted once the total emission load exceeds the total allowable load.

Article 6 of the Enforcement Ordinance indicates the permission for implementation plan for the TPLM with the contents below and Article 3 states the implementation assessment to be conducted every year.

- Current status of TPLM implementation plan areas
- Current status of pollution sources and prediction of future pollution sources
- Additional possible pollutant load from development projects (along with the detailed information of the development projects)
- Yearly target reduction load and measures
- Allocation of pollutant load
- Summary of data gathered from monitoring, water quality prediction and implementation

## 15.3.3 Reporting system of NPS management facility

Article 53 of the 'Act of Water Quality and Aqua-ecosystem Protection' states the reporting system on NPS pollution management facility to be applied in industrial areas and development projects listed in the Environmental Impact Assessment (EIA) program (Table 15.3). Developers of new construction projects and property owners of new manufacturing companies are required to install an NPS management facility in their properties to treat stormwater runoff. The installed facility must be reported and registered with the MOE as an environmental pollution treatment facility.

In case of existing companies, if there are certain property changes, as listed below, the property owners are obliged to install an NPS management facility and report it to the government.

- When a wastewater treatment plant is installed in the company (more than 10,000 $m^2$ property area)
- When the company has revised the EIA report
- When the property area is increased by more than 30% of the original area

## 15.3.4 NPS control area designation system

Article 54 of the 'Act of Water Quality and Aqua-ecosystem Protection' states that an NPS Control Area Designation System must be implemented to areas subjected to NPS pollution. Enforcement Ordinance indicates the criteria for the area designation as follows:

- Areas wherein more than 50% of the total load is contributed by NPS and water quality standards are not met
- Areas wherein the ecosystem can be extremely subjected to NPS pollutants
- Cities in which streams are highly impacted with NPS

- Industrial areas in which streams are highly impacted with NPS
- Areas requiring NPS management due to unusual geological and stratum features

**Table 15.3** List of manufacturing company and development projects requiring the installation of facility for NPS pollution management.

| Manufacturing Company | Development Projects |
|---|---|
| (1) Lumber and wooden products manufacturing | (1) Urban development project |
| (2) Pulp, papers and paper products manufacturing | (2) Industrial complex construction project |
| (3) Cokes, petroleum refinery products, and nuclear fuel manufacturing | (3) Energy development project |
| (4) Chemical compounds and chemical products manufacturing | (4) Harbor construction project |
| (5) Rubber and plastic goods manufacturing | (5) Road construction project |
| (6) Nonmetallic mineral goods manufacturing | (6) Water resources development project |
| (7) Primary metal industry | (7) Railway construction project |
| (8) Coal, crude oil and uranium mining industry | (8) Airport construction project |
| (9) Metal mining industry | (9) River use and development project |
| (10) Nonmetallic mineral mining industry | (10) Land and public waters reclamation project |
| (11) Beverage and food goods manufacturing | (11) Tourist facilities complex development project |
| (12) Electricity, gas and steam industry | (12) Forest development project |
| (13) Wholesale trade and commodities brokerage | (13) Special zone development project |
| (14) Sewage treatment, solid wastewater treatment and cleaning-related service industry | (14) Physical plant facility installation project |
| | (15) Solid wastes treatment facility installation project |
| | (16) National defense and military facilities installation project |
| | (17) Extraction project of soil/rock, sand, gravel, minerals, etc. |

### 15.3.5 Biotope and ecological area secure system

An ecologic environment wherein both people and nature coexist is needed due to the continuous urbanization and industrialization that has the potential to destroy the ecosystem. A biotope and ecological area ratio system has been developed for ecosystem conservation and restoration. Biotope, which is a combined term of *bios* (life) and *topos* (ground) in Greek, means a single colony, crowded community or habitat of plants and animals. It is the spatial boundaries of a biological community habitat that maintain the minimum natural ecosystems in urban development. The biotope and ecosystem area ratio can build a physical greenery network in the city and can restore damaged urban ecosystems. This system is considered as an important measure in industrial areas.

## 15.3.6 Buffer storage basin development system

The aqua-ecosystem is threatened because of environmental hazards such as fire, explosion, chemical leakage, etc. in industrial areas. Pollutants from toxic chemicals are usually emitted through chemical accidents, oil spills, and stormwater runoff and can be transported to water bodies through the sewer networks. A buffer storage basin is a disaster prevention facility that can protect the aqua-ecosystems from toxic chemicals and environmental disasters. It temporarily stores stormwater and removes hazardous and toxic chemicals in industrial sites.

## 15.3.7 Eco-industrial park (EIP)

Large amounts of energy and water are used in industrial sites. Industrial processes generate wastes, wastewaters, and industrial by-products. In addition, many environmental conflict factors arise when industrial sites are located adjacent to residential areas. An eco-industrial park (EIP) supports the coexistence of community and industrial complexes by fundamentally solving the environmental conflict factors (Kim, 2009). An EIP is an industrial complex that mimics the natural ecosystem through recycling and energy systems; thus, building a network that maximizes resource recovery and energy efficiency (Figure 15.2). An EIP basically reduces the environmental pollutants by means of an industrial ecology system.

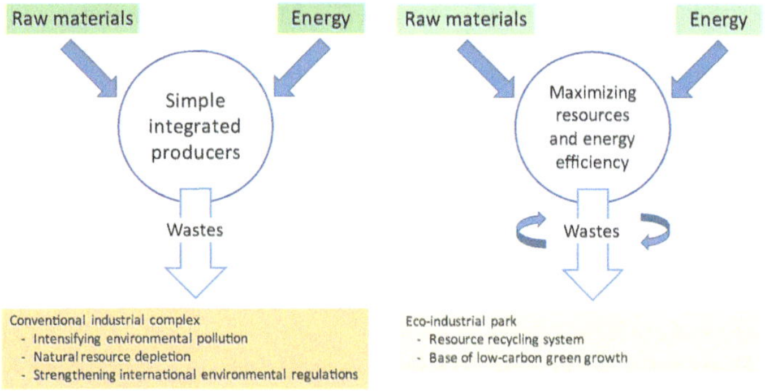


**Figure 15.2** Comparison of conventional industrial complex and EIP (EIP, 2011).

## 15.3.8 Integrated environmental management system

An industrial complex intensifies air pollution, water pollution, solid wastes, noise and vibration, odor, etc. Most developed countries have adopted a permit system to manage pollution. To comply with pollutant emission standards, treatment facilities should be installed with a different permit depending on the pollution type that needs to be managed. The Integrated Environmental Management System (IEMS) simplifies the permit system by applying an integrated management approach to manage pollution. The IEMS is based on the 'Act of Integrated Environmental

Management for Environmental Facility' which recommends the application of best available techniques (BAT) economically achievable to technically manage pollution.

### 15.3.9 Low impact development (LID) manual for EIA

Construction of urban areas and industrial complexes are development projects undermining the atmosphere, water, soil, ecosystems, etc. EIA is introduced to regulate the environmental impacts instigated by development projects. Recently, LID has been considered an important tool in the EIA system that helps to achieve a healthy water cycle and manage NPS pollution at the same time (Table 15.4). The LID manual has been developed to provide a systematic approach in applying LID in development projects. The manual includes an overview of LID, factors to be considered in LID application, design criteria of LID facility, etc.

**Table 15.4** Guidelines of LID technology.

| Content | Approach |
|---|---|
| Introduction of LID technology | – Benefits of LID application<br>– Introduction of related program/regulation |
| LID application measures at each development phase | – Basic investigation (related planning, watershed characteristics, hydraulics and hydrology investigation, soil and ground investigation, land use and detailed soil investigation)<br>– Check lists at each phase (phase of administration plan, phase of development plan)<br>– Check procedure for each development project (urban development, industrial development, road development) |
| Design and maintenance of LID technology | – Design considerations for each LID technology (installation criteria, plant selection criteria, maintenance criteria, etc.)<br>– Guidelines of installation and maintenance for each LID technology (bioretention, green roof, treebox filter, bio swale, bio slope, infiltration trench, dry well, porous pavement, sand filter, rain barrel) |
| Selection process of LID technology | – Basic principle of LID technology selection<br>– Selection of LID application location |

### 15.3.10 Restoration of water circulation and NPS management projects

LID techniques are widely applied in industrial sites and urban areas to restore the natural water cycle and manage NPS pollutants. However, the installation of a new treatment facility is difficult in an existing industrial park due to site restrictions. Many projects have been carried out in developed countries to find cost-effective approaches.

- *Water circulation city development project*: Aims to establish the water cycle and NPS treatment facilities in cities with a minimum of 100,000 population.

- *Green infrastructure development project*: Aims to apply green infrastructures to offices, schools, commercial sites, sports facility sites, etc. and provide technical education and publicity.
- *Civic participation project for NPS management*: Aims to engage citizens to treat NPS pollutants and restore the natural water cycle. As NPS management is essential in both public and private properties, civic organizations will perform various activities concerning NPS publicity and education, ordinance enactment initiatives, pollution monitoring, etc.
- *Ecological river development project*: Aims to create an ecological river aqua-ecosystem recovery and water quality improvement. The application of LID is essential in ecological river projects to provide stable and clean ecological waters in the watershed.

## 15.4 CONCLUSION

Industrial activities cause the discharge of pollutants to waterbodies. These pollutants such as organics, metals, nutrients, and toxic chemicals affect the water quality and ecosystem. Many developed countries have enacted regulatory regimes to manage industrial pollution. Programs include TPLMS, Reporting System of NPS Management Facility, NPS Control Area Designation System, Biotope and Ecological Area Secure System, Buffer Storage Basin Development System, EIP, IEMS, and LID Manual for EIA. However, the installation of a water cycle and NPS treatment facility is still challenging in an existing industrial park due to the lack of an available site. Thus, projects such as water circulation city development, green infrastructure development, civic participation, and ecological river development projects have been carried out in developed countries as cost-effective management approaches.

## REFERENCES

Eco Industrial Park (2011). Introduction to Eco Industrial Park. Retrieved from http://eip.or.kr

Kim K. H. (2009). A Potential Paradigm for Sustainable Regional Development: Eco-Industrial Park. Korea Institute for Industrial Economics & Trade. KIET Occasional Paper No. 74.

Kim L. H. (2003). Determination of event mean concentrations and first flush criteria in urban runoff. *Environmental Engineering Research*, **8**(4), 163–176.

Kim L. H., Ko S. O., Jeong S. and Yoon J. (2007). Characteristics of washed-off pollutants and dynamic EMCs in parking lots and bridges during a storm. *Science of the Total Environment*, **376**(1–3), 178–184.

Ministry of Environment (MOE) (2006). Guidance Manual to Control Nonpoint Source Pollution. Korea Ministry of Environment, Gwacheon.

Ministry of Environment (MOE) (2012). The Second Comprehensive Measures for NPS Pollution Management. 11-1480000-001222-01, Ministry of Environment. [Korean Literature]

Ministry of Environment (MOE) (2013). Source-Specific Management Measures: Nonpoint Source (NPS) Management Measures. Retrieved from http://eng.me.go.kr

Ministry of Environment (MOE) (2016). Water Pollution Load Management. Retrieved from http://eng.me.go.kr

United States Environmental Protection Agency (1999). Preliminary Data Summary of Urban Storm Water Best Management Practices. Office of Water, U.S. Environmental Protection Agency, Washington, D.C., EPA-821-2-99-012.

# Chapter 16

# Evaluating performance of proprietary and conventional urban stormwater management systems

*James H. Lenhart, PE, D.WRE*

*Contech Engineered Solutions LLC, Portland, Oregon, USA (E-mail: JLenhart@conteches.com)*

## 16.1 INTRODUCTION

In most developed nations agencies are struggling with the issues associated with stormwater management. Three principal elements of managing stormwater runoff are water quality, peak flow reduction and runoff volume reduction. Particularly in what has been termed the ultra-urban environment where land space and cost are high, there is demand for a variety of commercial stormwater treatment and management products available to developers for use in redevelopment and new build situations. They are also needed for public authorities and water utilities to apply in retrofit situations.

Processes for reviewing and approving these stormwater control measures are needed such that regulators have a high level of confidence that these stormwater control measures (SCMs) will function as intended and meet regulatory requirements for water quality, peak flow reduction and volume reduction. Unfortunately, regulatory treatment guidelines are often simplistic and create confusion as to how to evaluate data collected from performance studies.

The end result of this confusion can be misuse (or no use) of these SCMs even though their use is needed to meet both economic and engineering constraints. For example, in the US, the loosely defined term of Maximum Extent Practicable (MEP) serves as a guideline to agencies for approving and specifying SCMs. Though the intent of this guideline is relatively clear, the end result is often the

© IWA Publishing 2017. Wealth Creation without Pollution: Designing for Industry, Ecobusiness Parks and Industrial Estates
Brian D'Arcy, Lee-Hyung Kim, Marla Maniquiz-Redillas
doi: 10.2166/9781780408330_229

lowest cost (and typically least effective) Best Management Practice (BMP) and the result is the minimum extent practicable. This leads to the need to establish SCM evaluation and approval programs which rely on good data sets and are conducted through scientific method and peer review. The program also needs to set a bar and establish a level playing field for the manufacturers.

These products can offer significant benefits in terms of demonstrable effectiveness and cost saving but environmental regulators may be reluctant to accept them as sustainable drainage systems (SUDS; e.g., in the UK and Sweden) or valid WSUD (water sensitive urban design techniques in Australia) or LID (Low Impact Development stormwater features in USA, and elsewhere). Though there are emergent programs in Australia and New Zealand, in the USA a number of programs and rigorous protocols have been developed for verification of claims of manufacturers which eventually can lead to approvals by an agency for use within their jurisdiction. This paper sets out the basis for establishing an assessment and validation protocol, which can be an essential part of the regulatory regime required to ensure fit-for-purpose application of the technology.

## 16.2 CORE ISSUES

There are a number of core issues which lead to confusion. First is that we apply simple regulations and standards to complex issues. For example regulatory performance requirements use simple percent removals such as 80% Total Suspended Solids (TSS) or 50% Total Phosphorus (TP). However SCM performance is far more complex and requires methods of analysis far more sophisticated than comparison to a simple percent removal (Lenhart, 2007a, b). Also, many of the parameters measured and analytical methods performed are not wholly representative of stormwater constituents or do not include parameters such as particle size distribution which have a direct influence on how an SCM will actually perform.

Second, is that the regulator is the gatekeeper for making decisions on which SCMs to accept or reject. Unfortunately, in the light of reduced budgets and a relatively new applied science, many regulators are not well informed on how to make good decisions. Many agencies do not have the required processes or resources to methodically review and evaluate SCMs for use in their jurisdiction. Conceptually, agencies should be setting the bar for performance and establishing a level playing field for all SCM technology, however this is frequently not the case.

Lastly, in most engineering disciplines such as structural engineering or wastewater engineering, accountability is well defined and methods of analysis are more deterministic versus probabilistic. If a design does not meet performance expectations, it would be up to the engineer, contractor or manufacturer to fix, replace or 'make good' on design. However, in the stormwater treatment SCM market, there is little accountability with respect to proper design, construction and performance. Engineers commonly use cookbook designs that in reality may not perform as intended or are misapplied for the land use.

Once installed there is little or no follow up to ensure the SCM is functional and very few programs exist to monitor performance or require maintenance on a long-term basis. To a large extent, once a facility is completed, the site is fully permitted and the 'players' move to a different venue and the SCM is forgotten. This applies just as much too soft engineering ponds, swales, basins, etc. as it does to proprietary structures.

## 16.3 EXISTING PROGRAMS IN THE UNITED STATES

There are a number of state and local programs to review, verify, and approve manufactured SCMs in the US. Two of note are the New Jersey Corporation for Advanced Technology (NJCAT) verification program, and the State of Washington's TAPE (Technology Assessment Protocol Ecology) program.

The NJCAT program (www.NJCAT.org) evolved from an earlier program and now serves the State of New Jersey, though other States and local jurisdictions recognize the program. The original protocol was focused on both laboratory and field testing of SCMs. In the past couple of years the protocols were revised to be laboratory verification of products only. Laboratory testing protocols were developed using specific particle size distributions of silica sand for both hydrodynamic separators and filters. Hydrodynamic separator units can be used for pretreatment and filters can be used as standalone treatment. These protocols were developed in concert with SWEMA (Stormwater Equipment Manufacturers Association).

Basically a manufacturer must present a Quality Assurance Project Plan (QAPP) for review and approval. If accepted, the manufacturer moves to a test phase with an independent observer or laboratory. If not accepted, changes are negotiated with NJCAT. Once the testing is complete a draft finding report is published by NJCAT. This report is placed for public comment which includes other manufacturers. After the comment period, changes are addressed and resolved. Once the report is finalized it is issued as a verification and then can be used for certification by the State of New Jersey Department of Environmental Protection or other jurisdictions which recognize this process.

The Technology Assessment Protocol Ecology (TAPE) program was established by Washington State Department of Ecology (ECOLOGY) to address concerns that SCM performance was not actually meeting the State's criteria for TSS removal of 80%.

Having evolved over 15 years, a few iterations of the program have been developed based on knowledge gained from earlier programs. Currently, through State grants and participation fees, the Washington Stormwater Center (http://www.wastormwatercenter.org) was formed in 2009 to help facilitate the program for ECOLOGY. The center vets a board of external reviewers which provide peer review of manufacturer QAPPs and final Technical Evaluation Reports. Once reviewed, and deemed to meet the criteria and performance goals, a technology is granted the appropriate use designation.

To participate in the program a manufacturer must apply for one of three use designations each of which requires field testing which meets the TAPE protocol. A Pilot Level designation is assigned for laboratory scale based evidence which allows five sites to be installed for field monitoring to demonstrate it meets the performance goals. The second level is the Conditional Use Level Designation (CULD). To gain CULD the manufacturer must present both laboratory and substantive field data which may not meet the TAPE protocol criteria. If granted, then 10 sites are allowed for field monitoring. Lastly the General Use Level Designation allows for unlimited use of the technology assuming it meets both the protocol and the performance goals outlined in Table 16.1. Once granted GULD, ECOLOGY produces a certification letter which states the design criteria, maintenance guidelines, etc.

**Table 16.1** TAPE performance goals.

| Performance Goal | Influent Range | Criteria |
|---|---|---|
| Basic Treatment | 20–100 mg/L TSS | Effluent goal ≤ 20 mg/L TSS |
| | 100–200 mg/L TSS | ≥80% TSS removal |
| | >200 mg/L TSS | >80% TSS removal |
| Enhanced (Dissolved Metals) Treatment | Dissolved copper 0.005–0.02 mg/L | >30% dissolved Copper |
| | Dissolved zinc 0.02–0.3 mg/L | >60% dissolved Zinc |
| Phosphorus Treatment | Total phosphorus (TP) 0.1 to 0.5 mg/L | Must meet basic treatment goal and exhibit ≥ 50% TP removal |
| Oil Control | >10 mg/l | (1) No ongoing or recurring visible sheen in effluent<br>(2) Daily average effluent TPH concentration <10 mg/L<br>(3) Maximum effluent TPH concentration of 15 mg/L for a discrete (grab) sample |
| Pretreatment | <50–100 mg/L TSS | ≤50 mg/L TSS |
| | ≥100 mg/L TSS | ≥50% TSS removal |

*Source:* Courtesy Washington Department of Ecology.

The general process that must be followed is outlined in Figure 16.1. This method provides a level of independence and peer review to ensure confidence in the data and process and reporting. It also seeks to establish a common bar and a level competitive playing field for all technologies.

Approved technologies are then posted on ECOLOGY's website. (http://www.ecy.wa.gov/programs/wq/stormwater/newtech/technologies.html). Technologies approved by ECOLOGY can then be accepted by permitted jurisdictions. However, individual jurisdictions may choose not to accept the technology for use. The use of the product must also meet stated sizing requirements as well as designated maintenance requirements.

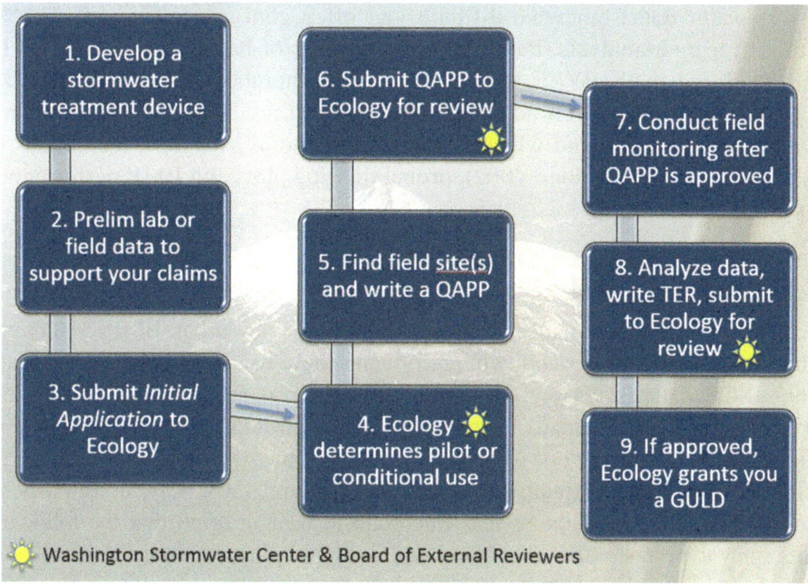


**Figure 16.1** Example of a stormwater accreditation process. *Source:* Courtesy Washington Department of Ecology.

More recent activity has included a committee investigation by the Water Environment Federation (WEF). The Stormwater Treatment Evaluation for Products and Practices (STEPP) (WEF, 2014) committee concluded there is a need for these products, and processes need to be in place to verify them for approval by various agencies. In concert with the United States Environmental Protection Agency (USEPA) and other organizations the STEPP committee is moving forward to establish a programmatic business plan.

## 16.4 METHODS FOR ASSESSMENT AND EVALUATING RESULTS

The validation of claims made for a given stormwater management technology is an essential element in the regulatory framework for bringing about effective stormwater management.

The development of QAPPs in relation to realistic performance goals set by the regulatory body are key to success. Field monitoring is a highly specialized discipline, and is time consuming and expensive.

Once the data are collected the reporting must be complete, and transparent with complete data sets and standardized report formatting.

Typically stormwater data are not normally distributed, they are highly variable and with a low population. These three factors make the analysis of stormwater data somewhat difficult and often confusing. There are no set methods for data analysis, but there are a number of basic tools which can be used. Outlined in the WEF Manual of Practice number 23 (WEF, 2012) are methods which have been adopted to review and analyze performance data. Of note are basic box and whisker plots for a general overview, techniques of linear regression (de Ridder, 2002), probability log plots and BMP performance expectation functions.

## 16.5 DISCUSSION

Uncertainty about performance is a characteristic of many urban BMPs/SUDS etc. features. High levels of variability in performance have been reported (e.g., Urbonas, 1999; ASCE database). How can variation in performance be reduced to more predictable levels? Avoidance of overload and misapplication is beyond the scope of this paper but is obviously one important factor. The other factors are concerned with design specifications and quality of build to make features which are fit for purpose. In considering risks of failures, it is important to bear in mind that the following issues apply to bespoke structures built by consulting engineers or others, as well as to proprietary units. Proprietary stormwater technology can offer the following benefits:

- Quality of build
- Elimination of uncertainty in performance if installed to manufacturer's specifications. Accountability by the manufacturer
- Technical validity of individual units, not just of an approach as applied by an individual developer following advice in a manual
- If small enough scale for plot by plot application, may reduce public sector liability
- Developers may be able to save on costs of additional measures (for example estate-scale flood risk management features, if they use plot by plot techniques).

One of the worries public regulatory bodies have in relation to proprietary systems is that they often have a requirement for more frequent maintenance than, for example, a sedimentation pond or wetland (smaller land-take means reduced storage capacity for sediment, which in turn must mean more frequent emptying of the accumulated sediment). The failure of land managers on industrial estates in the UK for example, to take out accumulated sediment and other pollutants regularly, has resulted in problems and consequently a reluctance to accept these techniques.

The same reluctance to undertake essential maintenance will of course eventually result in problems for all and any types of sustainable drainage or urban BMP types of structures, even ponds and basins. The problems can be effectively managed however for commercial units if a lease sales model is used, whereby the units are leased in effect by the customer from the manufacturer, and the lease payment includes an element for maintenance by the supplier or his approved contractor. Such an arrangement protects the reputation of the supplier and the product too. No such option is available for a pond or basin or other bespoke stormwater management landscape features.

## 16.6 CONCLUSIONS

- Clear and safe assessment protocols and independent validation of performance claims are essential for the introduction of reliable stormwater management techniques to the market.
- Regulators need to be confident about the performance of stormwater systems, whether, proprietary BMPs, permeable pavements, or other SUDS/ Urban SCMs/LID/WSUD features such as ponds, swales, basins, or bio filtration/retention features.
- It is important for commerce and international trade that comparable approvals standards are in place in the countries manufacturing and supplying stormwater management techniques.

## REFERENCES

De Ridder S., Darcy S. and Lenhart J. (2002). Influence of Analytical Method, Data Summarization Method, and Particle Size on Total Suspended Solids (TSS) Removal Efficiency. 9th International Conference on Urban Drainage. Report No. PD-02-006.1. Portland, OR.

Lenhart J. (2007a). Evaluating BMP's Programs, Success and Issues, Queensland Stormwater Industry Association. Annual State Conference. Sunshine Coast, Australia, Sept 2007 Keynote.

Lenhart J. (2007b). BMP Performance Expectation Functions – A Simple Method for Evaluating Stormwater Treatment BMP Performance Data. 9th Biennial Stormwater Research and Watershed Management Conference, April, 2007, Orlando, FL.

Schueler T. R. (1997). Comparative pollutant removal capability of urban BMPs: a reanalysis. Technical Note 95. *Watershed Protection Techniques*, **2**(4), 515–520.

Urbonas B. (1999). Design & selection guidance for structural BMPs. In: A. C. Rowney, P. Stahre and L. A. Roesner (eds), Sustaining Urban Water Resources in the 21st Century. ASCE, Reston, Virginia.

Water Environment Federation (2012). Design of Urban Stormwater Controls, WEF Manual of Practice No. 23, ASCE Manuals and Reports on Engineering Practice No. 87, WEF Press, Alexandria Virginia. Chapter 13, p. 513.

Water Environment Federation (2014). Investigation into the Feasibility of a National Testing and Evaluation Program for Stormwater Products and Practices, White Paper, downloadable: http://wefstormwaterinstitute.org/wp-content/uploads/2016/08/WEF-STEPP-White-Paper_Final_02-06-142.pdf

# Chapter 17

# An integrated approach for pollution prevention on industrial premises

*Brian J. D'Arcy*[1] *and Vicki Taylor*[2]

[1]*Independent environmental consultant (E-mail: brian@enviroexperience.co.uk)*

[2]*SEPA, Fife Office, Glenrothes*

## 17.1 EVOLUTION OF POLLUTION PREVENTION APPROACHES

Pollution risks from industrial premises have traditionally been identified in two broad categories:

- Aqueous waste streams which comprise an effluent discharge
- Spills and leaks of raw materials, wastes or products, which can contaminate drainage systems and the water environment directly, or via a public sewer connection.

Historically, efforts to reduce pollution from effluent discharges were focused on the provision of treatment plants with sufficient capacity to capture and remove problem pollutants and breakdown biodegradable ones. Frequently, trade effluents were discharged to a public sewer where persistent pollutants often adversely affected treatment processes or overloaded the public sewage treatment plants (Diamant, 1974). In the UK, initially in England and Wales with the formation of regional Water Authorities in 1974 and especially with the formation of privatised water utilities a decade later, the costs to the public sector of treating raw effluents from industry began to be put back to the industries, resulting in far more attention

© IWA Publishing 2017. Wealth Creation without Pollution: Designing for Industry, Ecobusiness Parks and Industrial Estates
Brian D'Arcy, Lee-Hyung Kim, Marla Maniquiz-Redillas
doi: 10.2166/9781780408330_237

being given to the sources within the industrial process plants. As industries faced significant costs for pre-treatment systems to reduce their effluent loadings and hence their trade effluent treatment charges, the value in the materials lost in the influent to a treatment plant began to be examined more critically (D'Arcy & Wither, 1984; Eckenfelder, 1989; D'Arcy *et al.* 1999). That trend is indicated in Figure 17.1.

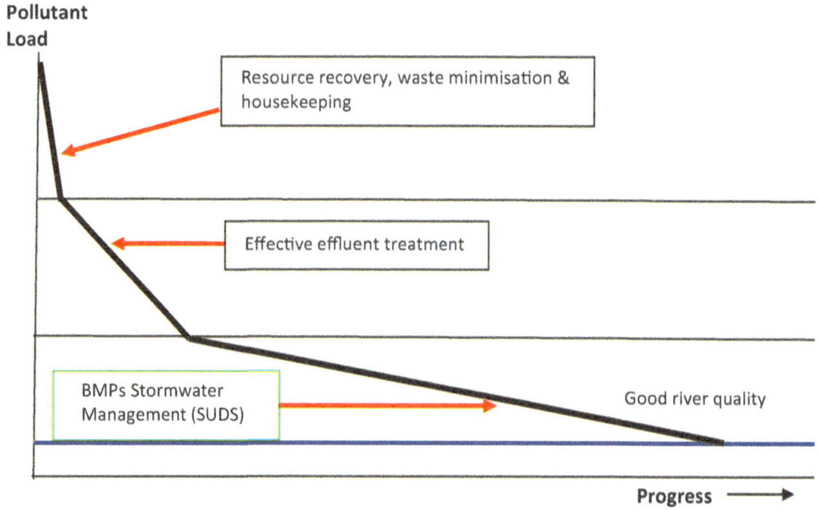


**Figure 17.1** Steps in achieving good river quality (BMPs: Best Management Practices; SUDS: Sustainable Urban Drainage Systems).

Figure 17.1 also shows a third stage in the progress towards greater efficiency and restoration of water quality; the implementation of urban BMPs or SUDS techniques to capture industrial diffuse pollution – the contamination of runoff from industrial premises (especially large complexes and estates). In such areas, contaminated surface water drainage can still prevent a watercourse reaching acceptable quality targets (D'Arcy & Bayes, 1995).

## 17.2 HOW TO ACHIEVE COMPLIANCE WITH EFFLUENT DISCHARGE LIMITS?

### 17.2.1 Process control

The key to compliance with a water quality discharge limit is knowledge of the sources of the raw effluent streams that are passing forward from production units in a factory for treatment at a wastewater treatment plant (WWTP) (Raak, 1987; Eckenfelder, 1989; D'Arcy, 1991). Figure 17.2 shows the variation in pollutant concentration in daily samples of the effluent from an industrial plant in Merseyside in the 1980s. Pressed to meet a tighter discharge standard, the company instigated a regime of daily sampling to allow identification of peak loads in influent streams to the plant. Obviously any WWTP has a design capacity and excessively high influent quality can overwhelm treatment capacity. Perhaps of greater long-term importance, high influent concentrations are associated with loss of resource from the business (Figure 17.2).

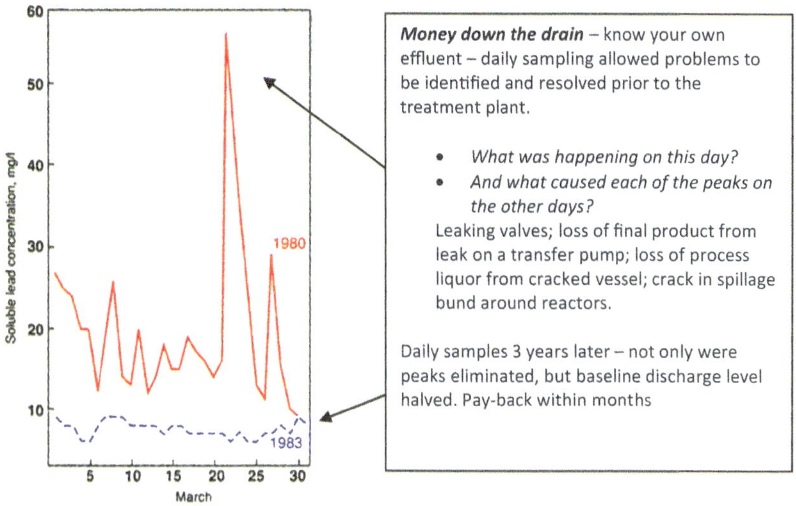


**Figure 17.2** Tracing the sources of influent variation in quality and flow, using daily samples and investigating the causes of identified peak concentrations and flows (adapted from D'Arcy, 1991).

As well as managing concentrations in drainage to a treatment plant, it is important to reduce hydraulic overload; mixing times in reactor vessels, and sedimentation rates, are susceptible to flow rate and the connection of large areas of impervious roof or yard and roads on an industrial site to the treatment plant is undesirable. Drainage arising from rainfall runoff does need consideration for treatment however; in the UK treatment systems have been licenced for stormwater discharges from a variety of industries from pharmaceutical chemicals, paper and board, to timber industries.

## 17.3 SITE RISK ASSESSMENT AND MANAGEMENT STRATEGY

Figure 17.3 sets out the key elements in a pollution prevention strategy for an industrial or commercial site. Important considerations for site operators can be grouped under three broad headings: drainage details; knowledge of all potential pollutants on site; and consequences, contacts and responsibilities. Each has implications and requires actions, as indicated below.

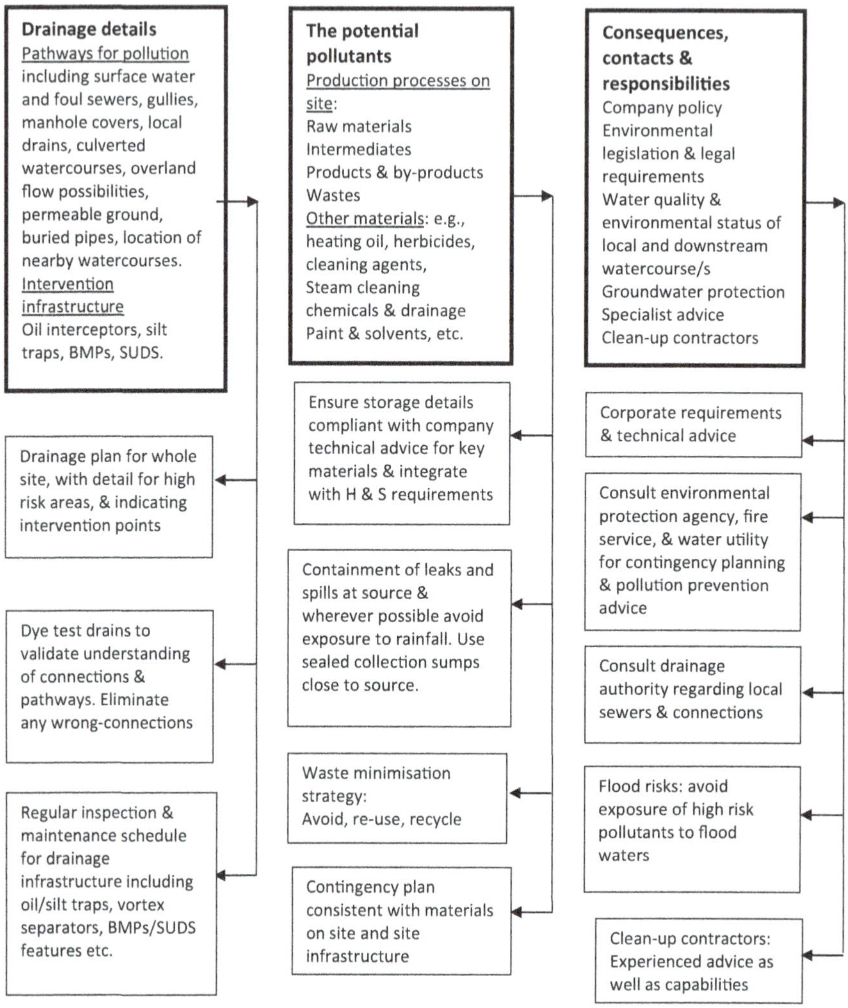


**Figure 17.3** Site pollution prevention planning: key considerations and information needs for a business: action chart for site manager.

The actions charted in Figure 17.3 need to be followed up by routine inspections and checks and a reporting procedure to site management.

## 17.3.1 Drainage details

A copy of a verified and clearly annotated Drainage Plan for the whole site, with detail for high risk areas, should be on the wall in the shift manager's office, with copies held by other key personnel.

As well as conventional drainage networks within a site, there are often additional pathways for pollution. Underground pipes in unmade ground, or under impervious cover that is old and cracked or subject to settlement, can be a chronic source of pollution. For example, oil feed pipes from an oil storage unit to a boiler house may corrode and leak without detection by operators for a long time, polluting groundwater and or finding entry to surface water systems. Repairs and renovation programmes should seek to replace such pipework with lines in an impervious pipe track, with open access for inspection and intervention. Access and inspection points on drains should be clearly identified on drainage plans, and contingency planning for accidents and emergencies needs to ensure that there is adequate access for road tankers and other clean-up equipment and crews. Similarly, access for routine maintenance of oil interceptors, silt traps, filter drains, grass filter strips and other BMPs or sustainable urban drainage features is essential. Any open systems such as grass swales have an essential role in cleaning up runoff, as well as managing major incidents, so should not be covered or replaced with pipes. Sites at risk of flooding need to consider potentially very different flow pathways and pollution risks under such circumstances (e.g., river flooding as well as pluvial floods).

### 17.3.2 Potential pollutants

Whatever the business, a waste minimisation philosophy will improve business efficiency and profitability and is the first action in developing a pollution prevention strategy. Waste generation is inefficiency and an increasingly expensive activity. Obviously the type of industry determines the character of the pollution risks on an industrial site, with pesticides, petrochemicals, many other chemicals, timber treatment and other industries presenting serious risks for stormwater management and accident contingency planning. But even low risk businesses are potential polluters if they have a vehicle fleet for distribution, loading, unloading, and especially when the vehicles are trucks and are maintained and steam cleaned on site (D'Arcy & Bayes, 1995). Detergent-oil mixtures can be more toxic than either substance on its own, and the drainage arising from such activities (a trade effluent) can be difficult to treat. A designated area drained either to the foul or combined sewer, or to a recycle facility is generally regarded as the best practice option.

Storage of wastes is often an issue on industrial and commercial sites (Figure 17.4). Sealed storage vessels for removal of wastes by approved contractors can deliver significant clean-up benefits on industrial yards. 'Empty' containers can be a source of pollution, for example when their original content was viscous oil which lines the sides and drains down slowly; if outdoors and exposed to rain the residual oil eventually floats to the surface and overflows or is spilled when enough drums accumulate for removal. Storage of empty containers as well as full ones, should be indoors or under a canopy to exclude rain, wherever possible, and drained to a sealed sump (see example guidance in Appendix).

**Figure 17.4** Empty drums and various wastes presenting pollution risks.

In contingency planning, consideration needs to be given to any chemical transformations, for example, involving products, packaging materials, and the fabric of the factory (paint, pipework etc.) that may occur in the heat of a fire. Planning for disposal of contaminated fire water is important: isolation valve/s on the internal drainage system, ideally on the periphery – perhaps a sump in a car park which can be flooded by firewater; agreements with local water utility and environmental protection agency on disposal of fire waters from the emergency storage (temporary) area/s on site.

### 17.3.3 Contacts, consequences and responsibilities

Environmental legislation and legal requirements are established in most countries, but many businesses are multi-national and may well have more stringent requirements to meet business reputation and corporate aims. Consultations with fire services and flood risk authorities need to be part of any contingency planning for a business on an industrial estate or business park where pollution risks are significant. A pesticides distribution depot, for example, may require careful contingency agreements with local farmers or foresters for disposal of fire water, and with drainage authority for intervention points to collect any. Such plans need careful discussion with the local environmental protection agency too, as well as the land owners and environmental heritage organisations. In at least one example, a large Swiss based company simply relocated one of its distribution depots from the shores of a RAMSAR wetland, to a coastal site after such discussions – both sites were within easy reach of customers. (The original premises were taken over by a plastic pipes distribution business; sustaining employment without the environmental risks). For some potential pollutants, collection and controlled disposal of fire water to foul sewer or an off-site waste management facility may be a preferred option.

The regulatory agencies and water utilities often have experience and expertise from dealing with many incidents over many years, and can provide specialist advice. Clean-up contractors sometimes also provide valuable experience and prevention guidance.

#### *17.3.3.1 Receiving water considerations*

Industrial premises exist in a catchment context, which has implications for pollution risk management. The water environment has several components, as indicated in Figure 17.5, and a variety of possible impacts need to be considered in a management strategy.

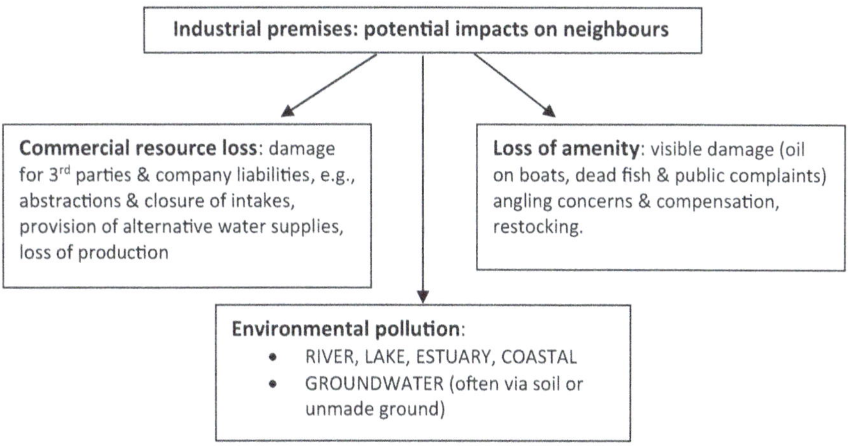


**Figure 17.5** Pollution incidents and possible impacts on responsible industry which require consideration in contingency planning.

Pollution risks to a watercourse may be obvious if the watercourse is alongside the factory, or close by. But many surface drainage networks are extensive and the receiving watercourse may be small and not obvious, in a culvert, or distant. Coastal locations may seem preferable in terms of dilution, and are in many circumstances, but local factors (e.g., a sailing club or sea angling) can certainly provide as much public concern from oil spills or other visible pollutants as may occur elsewhere. It is good neighbour practice for local businesses to join together in environmental best practice associations or local catchment projects.

Groundwater risks are important and long-term contamination often results from industrial activities, on unmade ground especially. In Alberta a campaign to address leaking underground storage tanks sought to address chronic pollution by hydrocarbons (Alberta, 1989).

The following sections deal in some detail with specific common pollution risk management designs and features for industrial and commercial premises.

## 17.4 OIL POLLUTION

### 17.4.1 Oil pollution risks at point of consumption or use

Oil incidents have consistently dominated pollution event statistics reported by the UK environment agencies, reflecting in part the greater visibility of oil films on watercourses, compared to the often-unseen chemical contamination. For example, Figure 17.6 compares numbers of oil and chemical incidents recorded by the Environment Agency (England and Wales) between 2004 and 2013.

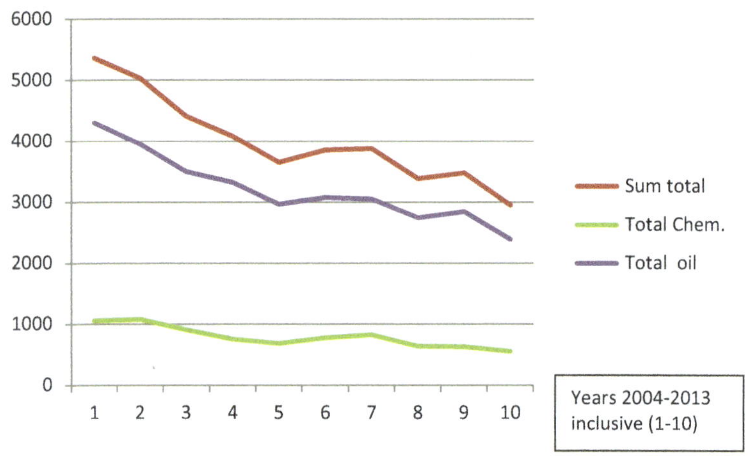


**Figure 17.6** Annual numbers of oil and chemical pollution incidents recorded by the Environment Agency (England and Wales, UK) over 10 years, 2004–2013 inclusive (EA, 2015).

The importance of oil as a ubiquitous diffuse pollutant is borne out by environmental studies such as the national urban streams sediment survey in Scotland (Wilson *et al.* 2005), and other studies have highlighted the primacy of industrial areas as diffuse pollution sources in urban contexts (Ellis & Chatfield, 2006).

### 17.4.2 Pollution sources for significant spills and leaks

Risk management in storage and handling oil, chemicals and other pollutants can be achieved by gathering as many of the risk points as possible into a single protected installation. The immiscible nature of oil with water, together with properties of relatively lower volatility for the higher molecular weight oils, allows for a broadly similar approach for risk management. Oil storage regulations have therefore been developed in many countries and are broadly similar in most features. Based upon annual returns from environmental regulators across the UK, a Department of the Environment (DoE) study provided the evidence for a government sponsored oil and chemicals working party to propose an ideal oil storage bunded installation The original raw data used is set out in Table 17.1, and shows that the majority of oil pollution incidents in the UK (inland waters) could be managed by putting the most regular causes of failures within a single bunded area (Figure 17.7).

**Table 17.1** The basis for oil regulations in UK. The 5-year dataset, 1986–1990, gives the distribution of oil incidents by source category.

| Source of Reported Oil Pollution | % of All Reported Incidents | Comments |
|---|---|---|
| Storage installations | 53% | Includes pump sets, failed tanks, over-filled tanks, fill points, open valves |
| Factory pipelines | 25% | Mainly associated with filter valves left off when pumping resumed |
| *Total containable in a bund* | *78%* | *If filters included within bund area* |
| Boiler/factory unit | 4% | Corrosion, contact & breakage, filters |
| Drums | 4% | Oil drums (full & 'empty') |
| Road traffic accidents | 4% | Leaking vehicle fuel tanks or loads |
| Other incidents | 13% | Miscellaneous causes. |

*Source:* Allcock *et al.* (1991).

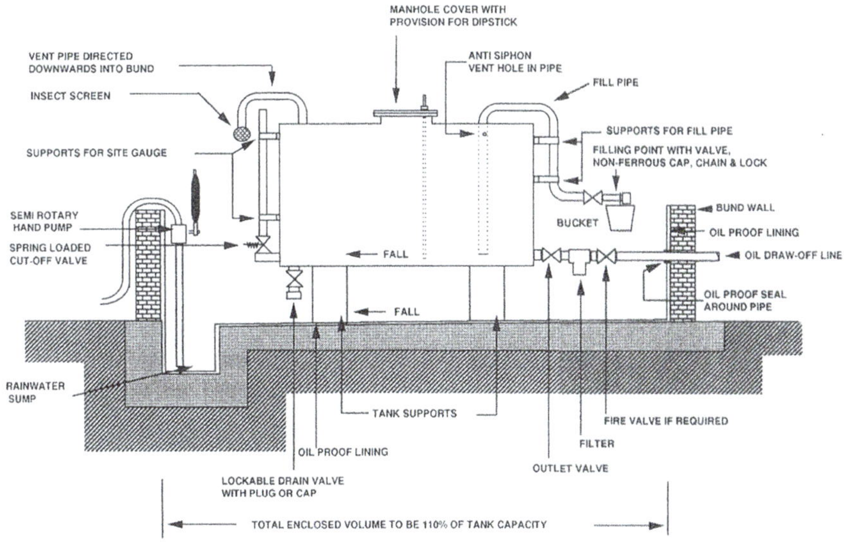


**Figure 17.7** Ideally bunded oil storage tank (Allcock *et al.* 1991).

Detailed causes of the incidents in Table 17.1 included: failures of storage tanks, leak from a bunded area, open drainage valve in a bund, overfilled tank, filler points outside bund, filter cover left off pipeline when pumping resumed, leaks at pump sets, oil drums, others, and unknown.

The design of an ideal bund was based on that (Figure 17.7) and has been taken into policy by all UK regulators (Ellis & Chatfield, 2006), and in the oil storage regulations that were eventually brought into force by subsequent legislation (Defra, 2001). An example installation in Laos is shown in Figure 17.8. The provision of a sump beneath the canopy for collection of leaks associated with pump sets would improve the installation in that example, and an increase in height of the bund wall, or fitting splash panels to divert spraying oil in the event of tank failures, would further improve pollution prevention. Unfortunately the filler points for the tanks in the bunded area are located outside the bund, creating a risk of uncontained spills and leaks (Figure 17.9).

**Figure 17.8** Oil storage installation showing inclusion of pump sets within a bunded area, with provision of a canopy to exclude rainfall.

For an idealised storage unit, with most risk points on the installation incorporated within the bund, the latter should be 110% of the volume contained in largest tank within the bund area. Bund wall height should be sufficient to capture spraying fluid under pressure (e.g., burst pipe or tank), or add a screen if wall and sump already meets 110% criterion but still too low for high pressure bursts.

In many countries, so-called integrally bunded storage tanks are available. It is important to recognise that such units may not be adequate if they have external connections for fill and use (see Figure 17.10).

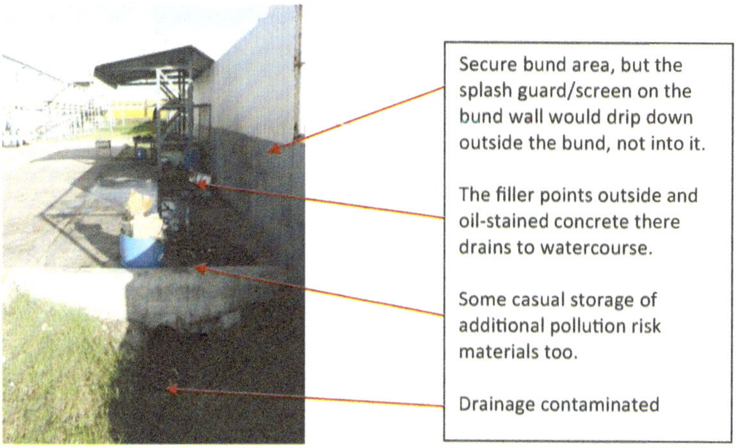


**Figure 17.9** Oil pollution risks created by locating fill points and storage outside the bund area.

**Figure 17.10** An integrally bunded tank that has filler points outside the bund and consequently still presents pollution risks.

### 17.4.3 Oil production, storage and distribution depots

The above advice refers to oil storage facilities at points of use within industrial sites. Large-scale storage installations that supply and distribute to the users' premises have to adopt a different approach. Pollution prevention is achieved by multiple checks and intervention points in the drainage networks serving the extensive area of outdoor bunded tanks. Drainage valves should be closed and padlocked, with drainage controlled by a supervisor with authority to open a valve and pass drainage through an interceptor, which may in turn pass forward to a site separator or treatment unit before final discharge. The eventual discharge should be licenced by the local regulatory agency, and discharged to a watercourse of sufficient size to provide adequate dilution for treated drainage, or discharged to a public sewer if sufficient hydraulic and treatment capacity, with approval of sewer authority.

## 17.5 CHEMICALS, TOXIC METALS AND OTHER POLLUTANTS

Pollution risks associated with chemicals are far more difficult to manage by general prescriptive measures of the type widely used for oil. For example, acids can be corrosive and require specially treated bund areas, strong acids and alkalis also present health risks and pools of potentially lethal material are undesirable. Other chemicals are sufficiently volatile to be a health hazard if allowed to spill into an open bund area. Protection of the water environment basically requires that *storage and handling areas should not drain into a surface water drainage system* (the simple basis of regulations in Scotland; discussed in D'Arcy *et al.* 2006). Detailed safety measures, including environmental pollution prevention, can then be determined by specific industries with the appropriate specialist knowledge of their chemicals and materials.

Pesticides storage is regulated in many countries, with statutory requirements to prescribe minimum pollution prevention measures. Detailed advice for storage and use is available from manufacturers and suppliers too, and voluntary trade associations also advocate best practices for pollution prevention.

Metal finishing and engineering industries are often characterised by elevated concentrations of toxic metals in yard or roof runoff (Maniquiz-Redillas & Kim, 2014). Site specific filter units, with removable filter bags to adsorb metal contaminants, are probably most applicable at such premises, with industrial connections for recovery of material or at least safe disposal. It is more sustainable to intercept at source, prior to mixing with contaminants from the broader industrial/urban drainage networks (Figure 17.11).

**Figure 17.11** Sources of day to day contamination by toxic metals, and BMP system for capture.

## 17.6 FOOD AND DRINKS INDUSTRY

Many substances which are not regarded as hazardous can also be highly polluting. For example, food industry products and raw materials such as milk, molasses, sugar, beer, grain, flour are all capable of polluting the water environment since spilled material, leaks and washing vans and yards after accidents, can all convey sufficient quantities of highly degradable material into surface water drainage systems and hence result in elevated BOD concentrations in receiving watercourses and anaerobic sediments. Three-figure BOD values have been measured for milk and beer for example, in stormwater drainage, and sources include leaks, spills and misconnections. Table 17.2 illustrates the potential for very significant contamination of industrial estate surface water drainage associated with food and drink industries.

**Table 17.2** Impacts of misconnections and other polluting activities on the quality of surface water drain discharging from Whitehouse industrial estate, Runcorn, UK, (annual mean values, 1975–1979) when breweries were being commissioned and new food industries were on site (unpublished data from North West Water Authority).

| Comments | Year | BOD mg/l | COD mg/l | Total Suspended Solids mg/l | BOD/SS Ratio |
|---|---|---|---|---|---|
| Significant pollution | 1975 | 60 | 237 | 186 | 0.3 |
| Soluble BOD dominating | 1976 | 200 | 289 | 87 | 2.3 |
| Soluble BOD still important | 1977 | 60 | 105 | 49 | 1.2 |
| 2 source areas diverted | 1978 | 26 | 105 | 92 | 0.3 |
| Still multiple minor source areas | 1979 | 37 | 160 | 128 | 0.3 |

Containment measures as set out above for oil installations can easily be adapted for such risks, allied with waste minimisation good practice. Additional measures may be appropriate on a site-by-site basis, following the general strategy set out above and taking into consideration individual characteristics of particular industries and processes. For example, grain at breweries and distilleries, bread-making and other factories is present in very large quantities and spilled grain is inevitable. Such material will settle in road drains on site and ferment, producing very high BOD drainage. Diversion of highest risk drains to foul sewer with agreement and subject to any requirements of the sewer authority may be an appropriate action, or preferably, incorporation into an energy recovery anaerobic digester. Sustainable drainage systems involving aerobic soil-grass features would be ideal for the rest of the site drainage.

## 17.7 DRAINAGE MISCONNECTIONS

Trade effluents that should not drain to surface water drainage systems are commonly found to be sources of pollution. Investigations at a large brewery in Merseyside identified trade effluent streams that were connected to the surface water drains rather than to the trade effluent treatment plant. Examples included waste beer from a beer-cans crushing plant for out-of-date beer, and some wash down areas (see Table 17.2). BOD concentrations can be higher than untreated sewage (D'Arcy *et al.* 1984). Mobile steam cleaning units are very common on many industrial premises and produce trade effluent that should be recycled or drained to foul sewer. Acidic drainage, for example, associated with a nickel plating plant in Dunfermline, Scotland, many years ago, corroded the concrete and resulted in severe pollution of a local watercourse. Sometimes a drain is simply connected to the wrong sewer in a dual system serving industrial premises. Sewage misconnections also occur on industrial premises, involving toilets, showers and canteens, for example. Dye testing drains can identify problems and allow simple restoration work. The possibility of misconnections from toilet blocks can be investigated by simple techniques such as temporarily inserting a ball of chicken wire tightly into surface water drain where it will trap sewage-specific debris for confirmation of a problem.

In some older factories, a building extension has resulted in rainwater downpipes being enclosed within the extended building, and open gullies therein provide pathways for pollution. Such access points for drainage to surface water systems inside buildings must be sealed off, since they cannot have any legitimate purpose. They might be suitable for reconstruction as sumps, but only if 100% sealed first.

## 17.8 WASTES

*Ad hoc* waste storage on external impervious surfaces creates a pathway for pollution (Figure 17.12). On a modern estate proper storage skips provide watertight containment to minimise leaks of pollutants and rainfall wash out.

**Figure 17.12** Uncontrolled waste storage on industrial yards, contrasted with waste storage units on a modern industrial estate: sealed containers, with no exposure to rainfall, and no drainage outlets.

## 17.9 CAPTURING POLLUTANTS WITHIN THE DRAINAGE SYSTEM

Sustainable urban drainage technology (D'Arcy, 2013; CIRIA, 2015) can provide day-to-day reduction in diffuse pollution, but also provides contingency planning advantages. Essential for new developments, retrofit opportunities may occur when sites are redeveloped, or as part of corporate clean-up actions, or as driven by pollution control regulators. Examples are given in Figure 17.13.

**Figure 17.13** Demarcation and collection drain for higher risk outdoor yard area (*left*). Rain garden taking road runoff and providing visible treatment facility with reduced risk of direct contamination of water environment in event of a spill (*right*).

## 17.10 CONTINGENCY PLANNING

If oil is stored on site in significant quantities it may be appropriate to hold absorbent material and floating oil booms on the premises as a stand-by measure (see Figure 17.14).

(a) (b) (c)

**Figure 17.14** Oil spill containment at industrial estate discharge to a retrofit SUDS pond in Livingston, Scotland and (a) contingency design in inlet to treatment pond from swale network *and* (b) penstock valve at outlet from first pond (c) a timber industries development, Lockerbie, Scotland.

Fires and floods, as well as leaks from storage vessels and road tanker accidents, must be part of a contingency plan.

- Know the drainage system and clearly mark key features on the ground, at the installations, to help emergency services in the event of incidents
- Keep up to date drainage plans where they can be easily accessed by a range of staff and emergency services in the event of incidents
- Plan for disposal of firewater and plan for storage and safe off-site disposal as may be necessary; review storage volume for fire water (3 $m^3$/tonne of stored material, for example, quoted by Salzmann (1987))
- Anticipate flood risks and prepare management responses
- Retain clean-up materials on site appropriate for all potential pollution incidents on the premises

Factory or laboratory fires are an important pollution risk (Salzman, 1987; Dowson, 1996; Abas *et al.* 2004). Detailed provision for fire water is an essential aspect of contingency planning, especially for high risk premises such as pesticides factories and distribution centres.

For flood risk premises, actions include removing toxic materials from lower level storage areas to elevated parts of warehouses, or to different premises in locations at greatly reduced risk of flooding, and relocation of pump-sets and power generators at higher levels within the premises. Flood measures include removable seals that can fit across doorways (including warehouse doors) and are more effective than sandbags.

## 17.11 DISCUSSION

Whilst there is a clear distinction between major spills and accidents, and the anthropogenic contamination evident as oil films and sediment from industrial estates, in reality there is a spectrum ranging between those extremes, of intermediate to minor contamination sources, often unavoidable, contaminating ground and runoff (Figure 17.15). In Scotland, the environmental protection agency, SEPA, is the responsible organisation for policing storage and handling requirements for oil, chemicals and other potential pollutants. The need to manage pollution from smaller-scale incidents mobilised by rainfall runoff (hence higher flows and lower concentrations) has to be addressed by the drainage facilities, and therefore once the runoff enters a public drainage system, the responsible body should logically be the water utility (in Scotland, Scottish Water). But the fact that there is no clear break point along the spectrum from major spills to almost continuous incidents and contamination, means that both agencies need to work together, and both should have *and use* appropriate regulatory powers.

The risks of higher concentrations of contaminants is greatest closest to source, and therefore source control SUDS should be present as well as housekeeping measures such as bunds and canopies over loading/unloading areas. That requires a change in focus for regulators and others engaging with industry and businesses: acceptance of diffuse pollution requirements to control many individually minor sources by actions at all such sources, rather than looking at each of the premises in isolation and assessing individual significance for a catchment.

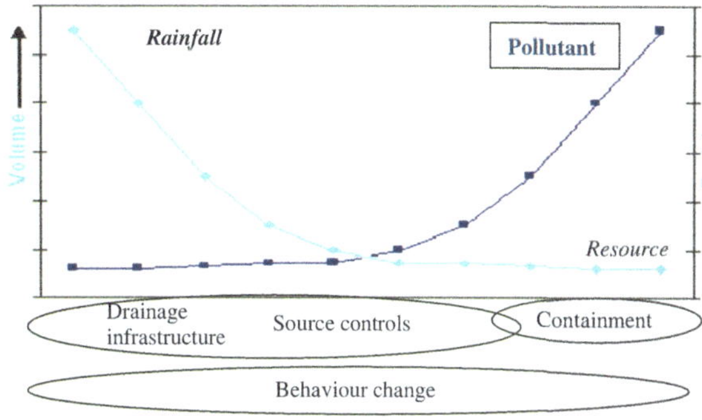


**Figure 17.15** Different pollution control approaches for diffuse sources of pollution, sorted by volume and toxicity (modified from Campbell *et al.* 2004).

Efforts to unequivocally indicate risks of pollution in surface water drains can present new risks. On sewer maps blue indicates surface drainage, and red indicates foul drains. Red however, (e.g., traffic lights, warnings generally), also often indicates a prohibition – a non-drainage specialist may assume therefore that the drain that should *not* receive the contaminated drainage. A shop floor operator or truck driver may interpret the information differently from a drainage engineer or regulator. A better practice has been applied with success in several countries: putting a fish or frog symbol alongside or onto the surface water drains. Perhaps manhole cover drains and grids could include such a feature, if configured to preclude misapplication over a foul drain; a need for water industry dialogue with suppliers (Figure 17.16)?

**Figure 17.16** Trying to prevent pollution from misconnections or casual disposal into the wrong drain.

A preferable option is to replace buried drainage systems, as redevelopment opportunities occur, with surface features such as gravel trenches or grass features at margins of industrial/commercial yard areas. These enable poor practices, such as disposal of waste oil, or leaks to be easily seen and resolved by management (Figure 17.17).

**Figure 17.17** Surface drainage using gravel trenches allows for easy detection of poor practice disposal and leaks. Also provides a first stage of degradation for lesser levels of contamination by microbes on the gravel.

## 17.12 CONCLUSIONS

A thorough assessment of pollution risks on industrial premises is the responsibility of the business operating on those premises. There are key areas of activity on which to focus, and effective tried and tested options for pollution prevention and clean up. Free advice and assistance can often be provided by business or trade associations, and by environment protection professionals. Pollution prevention measures may be cost neutral or even save money, if spills and leaks are contained and recovered close to source. As well as containment and contingency planning for major incidents, features to attenuate pollutants carried in runoff are desirable at a site, to contribute to catchment-scale water quality improvements. The greater ease of inspection of surface source control features facilitates good housekeeping assessment by management and staff. Inefficient use of resources, including failures to maximise value from materials on site, jeopardises long term sustainability of a business, as well as risking the availability of the receiving water as a good quality resource for neighbouring businesses and local communities.

## REFERENCES

Abas M. R. B., Omar N. Y. M. J. and Maah M. J. (2004). An assessment of the concentrations of particulate polycyclic aromatic hydrocarbons (PAHs) in the aftermath of a chemical store fire incident. *Journal of Environmental Sciences*, **16**(5), 751–754.

Alberta (1989). MUST Project – Management of Underground Storage Tanks. Alberta Environment, Edmonton, Canada.

Allcock R., D'Arcy B. J., Dolby J. and Walker R. (1991). Pollution of Water by Oil, First report of the Working Group on Pollution of Water by Oils and Chemicals, unpublished report. Water Environment Division, Department of the Environment, Marsham Street, London.

Campbell N., D'Arcy B., Frost A., Novotny V. and Sansom A. (2004). Diffuse Pollution: An Introduction to the Problems and Solutions. IWA Publishing, London. ISBN: 1 900222 53 1.

CIRIA (2015). The SuDS Manual (C753). CIRIA, London. http://www.ciria.org/Memberships/The_SuDs_Manual_C753_Chapters.aspx

D'Arcy B. J. (1991). Legislation and the control of dye-house pollution. *Journal of the Society of Dyers and Colourists*, **107**, 387–389.

D'Arcy B. J. and Bayes C. D. (1995). Industrial estates: a problem. In: C. Pratt (ed.), Proceedings of the Tenth Meeting of the Standing Conference on Stormwater Source Control. School of the Built Environment, Coventry University, Coventry. ISBN 0 905949 33 1.

D'Arcy B. J. and Wither A. W. (1984). Reductions in Effluent Loads Achieved during a Period of Economic Recession. Unpublished technical report of North West Water ref. no. RI 84/1 (original paper given to the Industrial Water Society, 11th October 1983), Warrington.

D'Arcy B. J., Wilbraham D. J. and Williams B. (1984). Pollution from Industrial Estates. NW Water Authority unpublished technical report RI84/2.

D'Arcy B. J., Schmulian K. and Wade R. (2006). Regulatory options for the management of rural diffuse pollution. In: L. Gairns, K. Crighton and B. Jeffrey (eds), Managing Rural Diffuse Pollution. Proceedings of the SAC & SEPA Biennial Conference, 5–6 April 2006, Edinburgh, pp. 192–200, ISBN 1-901322-63-7.

D'Arcy B. J., Todd R. B. and Wither A. W. (1999). Industrial effluent control and waste minimisation: case studies by UK regulators. *Water Science and Technology*, **39**(10), 281–287.

Defra (2001). Water Resources, England. The Control of Pollution (Oil Storage) (England) Regulations 2001. HMSO, London.

Diamont R. M. E. (1974). The Prevention of Pollution. Pitman Publishing, New York and London, ISBN 0 273 31819 5.

Dowson P., Scrimshaw M. D., Nasir J. M., Bubb J. N. and Lester J. N. (1996). The environmental impact of a chemical spill from a timber treatment works on a Lowland River System. *Journal of Chartered Institution of Water and Environmental Management*, **10**, 235–244.

Eckenfelder W. W. (1989). Industrial Water Pollution Control, 2nd edn. McGraw-Hill Book Co., London. ISBN 0-07-018903-X.

Ellis J. B. and Chatfield P. R. (2006). Diffuse urban oil pollution in the UK. *Urban Water Journal*, **3**(3), 165–173. September 2006.

Environment Agency (2015). Freedom of Information Act enquiry. Environment Agency Information © Environment Agency and database right. www.environment-agency.gov.uk

Maniquiz-Redillas M. C. and Kim L.-H. (2014). Fractionation of heavy metals in runoff and discharge of a stormwater management system and its implications for treatment. *Journal of Environmental Sciences*, **26**, 1214–1222.

Raak Y. (1987). The fight against industrial and urban pollution in the Artois-Picardie basin. *Water Science & Technology*, **19**, 371–379.

Salzmann J. J. (1987). "Schweizerhalle" and its consequences. In: Fires Involving Pesticides – The Rhine Incident. Proceedings of a Seminar on 25/26 November 1987. British Agrochemicals Association.

Wilson C., Clarke R., D'Arcy B. J., Heal K. V. and Wright P. W. (2005). Persistent pollutants urban rivers sediment survey: implications for pollution control. *Water Science & Technology*, **51**(3–4), 217–224.

# APPENDIX

Example pollution prevention guidance and sources of further information (UK housekeeping advice, see elsewhere for suds).

| Organisation | Publication | Date | Weblink | Comments |
|---|---|---|---|---|
| EA | PPG 1: Understanding your environmental responsibilities – good environmental practices | July 2013 | http://www.sepa.org.uk/media/60060/ppg-1-general-guide-to-the-prevention-of-pollution.pdf | Good source of information on good practice, not as a source of legal requirements |
| EA | PPG 2: Above ground oil storage tanks | August 2011 | http://www.sepa.org.uk/media/60073/ppg-2-above-ground-oil-storage.pdf | Good source of information on good practice, not as a source of legal requirements |
| EA | PPG 3: Use and design of oil separators in surface water drainage systems | April 2006 | http://www.sepa.org.uk/media/60086/ppg-3-use-and-design-of-oil-separators-in-surface-water-drainage-systems.pdf | Good source of information on good practice, not as a source of legal requirements |
| EA | PPG 4: Treatment and disposal of sewage where no foul sewer is available | July 2006 | http://www.sepa.org.uk/media/60099/ppg-4-treatment-and-disposal-of-sewage-where-no-foul-sewer-is-available.pdf | Good source of information on good practice, not as a source of legal requirements |
| EA | PPG 5: Works and maintenance in or near water | October 2007 | http://www.sepa.org.uk/media/60112/ppg-5-works-and-maintenance-in-or-near-water.pdf | Good source of information on good practice, not as a source of legal requirements |
| EA | PPG 6: Working at construction and demolition sites | 2012 | http://www.sepa.org.uk/media/60125/ppg-6-working-at-construction-and-demolition-sites.pdf | Good source of information on good practice, not as a source of legal requirements |
| EA | PPG 7: Safe storage – The safe operation of refuelling facilities | July 2011 | http://www.sepa.org.uk/media/60138/ppg-7-safe-storage-the-safe-operation-of-refuelling-facilities.pdf | Good source of information on good practice, not as a source of legal requirements |
| SEPA | SEPA: 'Underground storage tanks for liquid hydrocarbons: code of practice for the owners and operators of underground storage tanks (and pipelines)' | May 2006 | http://www.sepa.org.uk/media/145052/code_of_practice_for_installers_owners_and_operators_of_underground_storage_tanks_pipelines_water.pdf | Based on Controlled Activities Regulations 2005 |

| | | | | |
|---|---|---|---|---|
| EA | PPG 8: Safe storage and disposal of used oils | February 2004 | http://www.sepa.org.uk/media/60151/ppg-8-safe-storage-and-disposal-of-used-oils.pdf | Good source of information on good practice, not as a source of legal requirements |
| EA | PPG 13: Vehicle washing and cleaning | July 2007 | http://www.sepa.org.uk/media/60164/ppg-13-vehicle-washing-and-cleaning.pdf | Good source of information on good practice, not as a source of legal requirements |
| EA | PPG 21: Pollution incident response planning | March 2009 | http://www.sepa.org.uk/media/100557/ppg-21-pollution-incident-response-planning.pdf | Good source of information on good practice, not as a source of legal requirements |
| EA | PPG 22: Incident response – dealing with spills | April 2011 | http://www.sepa.org.uk/media/60177/ppg-22-incident-response-dealing-with-spills.pdf | Good source of information on good practice, not as a source of legal requirements |
| EA | PPG 26: Safe storage – drums and intermediate bulk containers | March 2011 | http://www.sepa.org.uk/media/60190/ppg-26-safe-storage-drums-and-intermediate-bulk-containers.pdf | Good source of information on good practice, not as a source of legal requirements |
| EA | PPG 27: Installation, decommissioning and removal of underground storage tanks | April 2002 | http://www.sepa.org.uk/media/100570/ppg-27-installation-decommissioning-and-removal-of-underground-storage-tanks.pdf | Good source of information on good practice, not as a source of legal requirements |
| NIEA & SEPA | NetRegs – Environmental Guidance for your Business | | http://www.netregs.org.uk/ | Access to guidance, legislations and e-learning tools |
| EU | Zero Waste Europe | | https://www.zerowasteeurope.eu | Reports and case studies on waste reduction |
| EA | Waste duty of care code of practice | March 2016 | https://www.gov.uk/government/publications/waste-duty-of-care-code-of-practice | Practical guidance on meeting duty of care requirements (England and Wales) |
| EA | Is your site right? 10-point checklist | | http://www.sepa.org.uk/media/100518/is_your_site_right.pdf | Checklist for oil and chemical storage, waste management and site drainage |

(*Continued*)

Example pollution prevention guidance and sources of further information (UK housekeeping advice, see elsewhere for suds) (*Continued*).

| Organisation | Publication | Date | Weblink | Comments |
|---|---|---|---|---|
| EA | Pollution Prevention Pays – Getting your site right | January 2013 | https://www.gov.uk/government/uploads/system/uploads/attachment_data/file/485173/LIT_7481.pdf | Guide to preventing industrial and commercial pollution |
| WRAP | The Waste and Resources Action Programme | | http://www.wrap.org.uk/ | Practical environmental support for business |
| CIRIA | Containment systems for the prevention of pollution | July 2014 | http://www.ciria.org/Resources/Free_publications/c736.aspx | to assist owners and operators of industrial and commercial facilities storing substances (inventories) that may be hazardous to the environment. |
| CIRIA | Above-ground proprietary prefabricated oil storage tank systems (C535) | April 2002 | http://www.ciria.org/ItemDetail?iProductCode=C535&Category=BOOK&WebsiteKey=3f18c87a-d62b-4eca-8ef4-9b09309c1c91 | Assesses environmental protection offered by such systems |
| CIRIA | Chemical storage tank systems – good practice. Guidance on design, manufacture, installation, operation, inspection and maintenance (C598) | September 2003 | http://www.ciria.org/ItemDetail?iProductCode=C598&Category=BOOK&WebsiteKey=3f18c87a-d62b-4eca-8ef4-9b09309c1c91 | Advice on chemical storage good practice |
| CIRIA | Environmental good practice on site pocket book (fourth edition) (C762) | April 2016 | http://www.ciria.org/ItemDetail?iProductCode=C762&Category=BOOK&WebsiteKey=3f18c87a-d62b-4eca-8ef4-9b09309c1c91 | Advice on management of on-site environmental issues |

# Chapter 18

# Product substitution – addressing the challenge of hazardous priority pollutants

*J. B. Ellis[1]* and B. J. D'Arcy[2]*

*[1]Urban Pollution Research Centre, Middlesex University*

*[2]Independent*

**Corresponding author: bryan@chantelet.net*

## 18.1 INTRODUCTION

Product replacement can be considered as a primary risk management option for limiting environmental contamination by persistent, bioaccumulative priority pollutants (PPs). Such substitution needs to be considered on the basis of the cost-effectiveness of reducing and managing the risks associated with a particular chemical to an acceptable 'safe' level. The risk of the used substance must be compared with the risk of the substitute replacement product and this requires reliable data on the hazards presented by the alternative; for example, replacement of chlorinated hydrocarbons with organophosphate pesticides, or permethrins, or replacement of toxic metals in automobile engines and brake systems, with ceramic units. The European Union (EU) Water Framework Directive (WFD) has identified 13 PPs as being particularly hazardous and has deemed that they should be phased-out from all discharges and emissions within a 20-year timescale (European Commission, 2012). Table 18.1 lists the substances and indicates their sources, properties and environmental quality standards (EQSs). A further eight priority pollutants have been placed on a Watch List for possible future hazardous designation including glyphosate, mecoprop and Bisphenol-A, all of which have been commonly detected in diffuse rural and urban runoff. The reduction and

© IWA Publishing 2017. Wealth Creation without Pollution: Designing for Industry, Ecobusiness Parks and Industrial Estates
Brian D'Arcy, Lee-Hyung Kim, Marla Maniquiz-Redillas
doi: 10.2166/9781780408330_259

**Table 18.1.** Priority Hazardous Pollutants in the European Union (under the EU Water Framework Directive).

| Substance | Sources | Key Properties | | | Inland Waters EQS Standard (μg/l) | | Inland Waters Sediment Standard (mg/kg) | |
|---|---|---|---|---|---|---|---|---|
| | | **Half-life ($T_{0.5}$ days) in Surface Waters** | **Toxicity** | **BCF** | **AA-EQS** | **MAC-EQS** | **TEL** | **PEL** |
| **Anthracene** (PAH) | Combustion product, dyes, textiles, plastics, insecticides | High-592; Low-21 | Toxic | 532.7 | 0.1 | 0.4 | 0.1+ | |
| **Cadmium (+ compounds)** (Metal) | Electroplating, coatings, batteries, solar panels, pigments, brake pads/ tyres, solder, alloys, fertilisers, cement, tobacco smoke | – | Toxic, carcinogen, endocrine disrupter | 3.162 | 0.08–0.25 | 0.45–1.5 | 0.596 | 3.53 |
| **Chloroalkanes** (Paraffin HC) | Solvents, aerosols | High-4 yrs; Low-2 hrs | Non-biodegradable | 1173–>7000 | 0.4 | 1.4 | | |
| **Endosulfan** (Organochloride) | Insecticide | High-9; Low-4.5 hrs | Toxic, bioaccumulative, endocrine disrupter | 117.5 | 0.005 | 0.01 | | |
| **Hexachlorobenzene*** (Organochloride) | Fungicide/pesticide, dyes, wood preservative | High-2081; Low-986 | Carcinogen, bioaccumulative | 5153 | 0.01 | 0.05 | | |

| | | | | | | | | |
|---|---|---|---|---|---|---|---|---|
| **Hexachlorobutadiene*** (Chlorinated HC) | Solvents, lubricants, rubber production | High-183; Low-28 | Toxic, carcinogen | 956.3 | 0.1 | 0.6 | | |
| **Hexachlorocyclohexane*** (Organochloride) | Pesticides | High-135; Low-14 | Toxic | 307.5 | 0.02 | 0.04 | | |
| **Mercury (+ compounds)*** (Metal) | Electronics, fluorescent lamps, dentistry, cosmetics | – | Toxic, bioaccumulative | 69.34 | 0.05 | 0.07 | 0.174 | 0.486 |
| **Nonylphenols** (Organic alkylphenol) | Antioxidants, lubricants, detergents/soaps, emulsifier, pesticides, plastics, personal care products | High-50; Low-15 hrs | Bioaccumulative, endocrine disrupter | 543.5 | 0.3 | 2.0 | 1.4# | |
| **PAHs** (Organic HCs) | Major combustion product, oil leakage/ spills | High-12.5; Low-1 hr | Toxic, carcinogens, mutagenic, bioaccumulative | 7.83–244 | 0.002–0.03 | 0.1 | 31.7 – 41.9 | 385 –862 |
| **Pentachlorobenzene** (Chlorinated HC) | Fungicides, fire retardant | High-345; Low-194 | Toxic, bioaccumulative | – | 0.007 | N/A | | |
| **Pentabromodiphenolether*** (Organic POP) | Flame retardant | High-4.5; Low-1 hr | Toxic, bioaccumulative | | 0.4 | 1.0 | | |
| **Tributyltin (+ compounds)** (Organotin) | Biocide, fungicide, disinfectant | High-150; Low-6 | Toxic, bioaccumulative, endocrine disrupter | – | 0.0002 | 0.0015 | 0.1+ | |

*Notes:* *Much stricter EQS standards have been recommended by UKTAG (2013) for future EU river basin management.
+CEFAS (1994) Action Level 1 standard for river/estuarine dredged sediment; #Canadian sediment quality guidelines (SQGs);
PEL – Probable Effects Level (Time-weighted average); TEL – Threshold Effects Level (Time-weighted average); AA-EQS – Annual Average Environmental Quality Standard; MAC-EQS – Maximum Admissible Concentration; BCF – Bioaccumulation Factor; PAHs – Polyaromatic hydrocarbons; HC – Hydrocarbons; POP – Persistent organic pollutants.

eventual removal of such pollutants from surface water discharges to receiving waters presents a considerable challenge for stormwater management.

Structural end-of-pipe and source controls can collectively offer a variable level of capture and removal from stormwater discharges. This might be achieved as part of bringing into routine use stormwater management techniques such as best management practices (BMPs) or sustainable urban drainage systems (SUDS). But that approach leaves water utilities to pay the cost of disposal of the captured pollutants (as contaminated sludge in the BMPs/SUDS). A more effective, fairer and in the medium to long term more sustainable approach, would be to control these substances prior to emission, by product substitution. However, it must be recognised that this objective would be difficult to achieve for all the substances listed in Table 18.1. Approaches to product control and substitution take time of course, since they require the working consent of the industries involved, and possibly their customers too. In addition, total 'in-factory' pollution control may be beyond the technological or economic capability of many industrial/commercial groups and enterprises.

As all chemicals have specific properties, product replacement is not a straightforward issue as some PPs are derived unintentionally as unavoidable by-products in combustion and industrial processes. In general, they also have multiple sources varying from individual households to large manufacturing plants and regional diffuse sources, for example traffic, agricultural fertilisers/pesticides etc.

This makes effective targeting of product control measures both complex and politically sensitive (Mikkelsen *et al.* 2008). From a practical viewpoint, national regulatory and legislative authorities have to decide upon the most appropriate solution to a given pollution problem and this requires an assessment of the technical feasibility and performance effectiveness of differing options evaluated on the basis of clearly defined criteria. As a working principle, cost should never be the limiting factor where human health impacts should be paramount (see Harada, 1995), but may require the adoption of interim actions, and the timescale by which ultimately desirable controls can be achieved.

Product substitution is also highly dependent on the product process and usage, the risk level posed to economic and environmental health as well as the social costs and benefits. The decision to remove lead from petrol in 1973 was supported by political, commercial, public and environmental groups being firmly based on a strong environmental evidence base, especially human health impacts (Royal Commission, 1983), with clear national motivation and impact awareness. Over the following three decades, event mean concentrations (EMCs) of lead in UK urban runoff have been reduced by an order of magnitude from 1000 μg/l in the 1970s to 60 μg/l by 2008, with a lower 25% quartile level at 29 μg/l (Mitchell, 2001). For highway runoff in the UK, the current EMC lead level is only some 17% of the long term 1970–2000 average. However, many other potential alternative replacement products have been subject to considerable market resistance. The recent decision in the US to designate the organic DINP phthalate compound to replace the general DEHP plasticizer for PVC products has been hotly contested with studies alleging that the replacement

DINP possesses carcinogenic properties (although at much reduced environmental side effects), being labelled as 'misleading' and unsupported by the available weight-of-evidence (Office of Environmental Health Hazard Assessment [OEHHA], 2013).

Figure 18.1 provides a schematic outline of the stages involved in identifying and evaluating potential substitution products for PPs and their mitigating solutions. The first stage in the procedure requires the identification of hazardous substances currently in use and possible alternatives for the processes or products. Stage II considers alternative options based on technical and toxicological criteria with the Stage III decision guided by economic analysis.

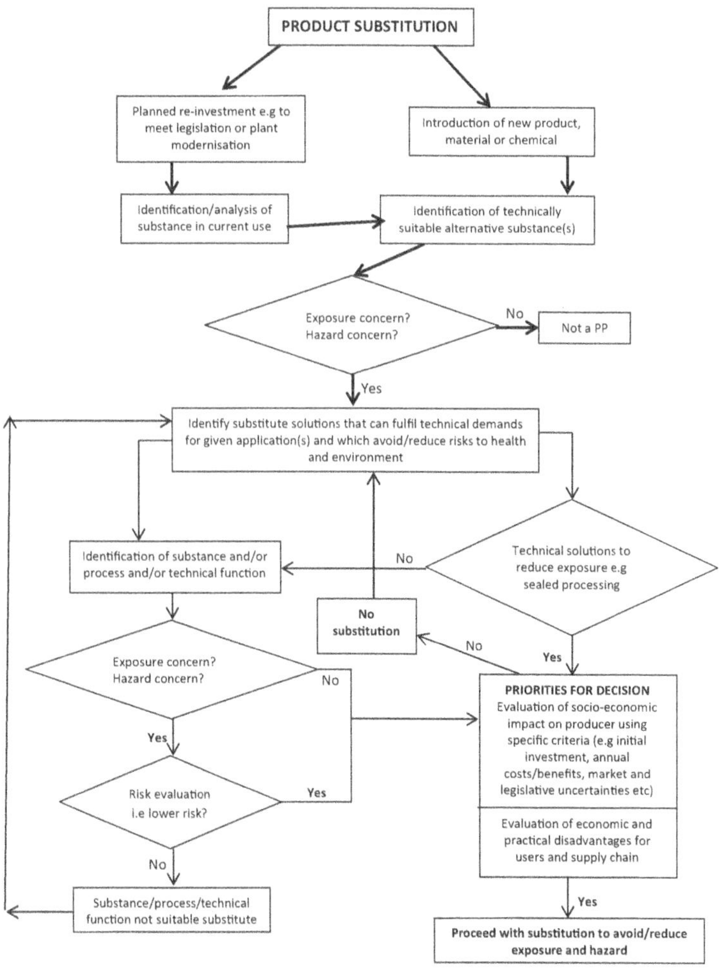


**Figure 18.1** Schematic flow chart for product substitution decisions.

Following any replacement implementation, there should be a continuous Stage IV follow-up and feedback analysis to ensure an effective safe performance without downstream user disruption. Proprietary safeguards in respect of material composition and testing evaluations often make this industrial source control substitution pathway a difficult, costly and prolonged route and one which might only be applicable for direct emissions to receiving waters (UK Chemicals Stakeholder Forum [UKCSF], 2010).

## 18.2 CADMIUM CASE STUDY

The search of suitable substitutes under the Stage I review should be initially focused on the uses leading to the most significant releases into the environment, and data underpinning this 'fit-to-use' approach for cadmium (Cd) is illustrated in Figure 18.2. It is estimated that total environmental emissions of cadmium in the UK are in the order of 23 tonnes per annum with 6.5 tonnes deriving from vehicular sources (OSPAR, 2004). In the UK, diffuse sources such as water supply and domestic activities account for 37% and 32% respectively of cadmium source contributions to sewer discharges with highway runoff contributing a further 17%.

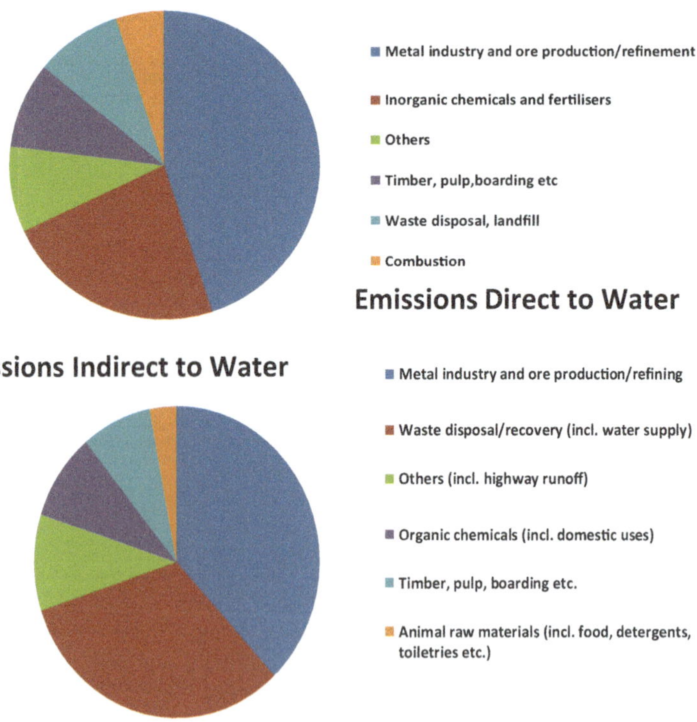


**Figure 18.2** Cadmium emissions to European freshwaters (E-PRTR, 2008).

Industrial effluent is estimated to contribute only 15% of the total sewer cadmium loading mainly related to electroplating, textile dyeing, battery and printing industries. Revitt *et al.* (2013) estimate that full implementation of the amended EU PP directive is likely only to achieve a maximum 50% reduction of emissions from these sources. In this instance it is difficult to argue that hard targeting of the manufacturing industry would yield justifiable benefits, particularly given that the majority (>70%) of cadmium releases are as dissolved species (European Pollutant Release and Transfer Register [E-PRTR], 2008), with secondary activated sludge (AS) treatment averaging 50% to 60% removal rates falling to an average of less than 20% when total suspended solids (TSS) influent concentrations to sewage treatment works (STWs) are low.

The current cadmium MAC-EQS is set at 5 μg/l but it is proposed under the amended EU priority substance directive (European Commission, 2012) to reduce this for all inland waters to the equivalent US continuous critical concentration (CCC) of 0.5 μg/l (for the same hardness values). The best and worst final effluent cadmium concentrations for some 30 UK STWs have been estimated as being 0.08 and 0.6 μg/l respectively (Ziolko *et al.* 2011). Thus whilst optimised conventional sewage treatment might go some way to meet the proposed upgraded cadmium standards, compliance would be very dependent on local effluent, site and operational conditions. Both Ziolko *et al.* (2011) and Revitt *et al.* (2013) consider that advanced wastewater treatment technologies might substantially reduce direct cadmium discharges to receiving waters although leaving indirect discharges largely unaddressed.

There has already been a considerable replacement policy for cadmium in product uses such as battery technology, soldering alloys, paints and coatings. Exact tracing and understanding of sources is currently insufficient to be reliably and efficiently used to direct and manage effective emission control strategies. Control of small enterprises such as dentists or car washing would probably contribute significantly to reduce Cd and other metal discharges to sewers. However, reduction for these specific sources is beyond the management control of wastewater companies and depends on the motivations and actions of wider societal and political groups. Revitt *et al.* (2013) have shown that direct cadmium emissions to groundwater and soil dominate under all emission control strategies with the lowest emissions being to sludge, sediment and air but with voluntary mitigation measures and source best available technology (BAT) providing the most effective product control strategies. Stormwater BMPs and source control strategies were estimated to offer cost-effective mitigation but were partly offset by long-term cadmium accumulation in sludge and sediment. Unfortunately, cadmium cannot always be avoided as a by-product in many industrial processes and substitutes for some uses such as nickel-cadmium batteries or cadmium telluride materials (e.g., in sensors, switches, solar cells etc.) are yet to be developed.

Wherever feasible, cadmium products should be recycled with incentive and information campaigns developed for all stages of the consumer chain, particularly for those areas where substitution is not currently possible. A more effective mitigation strategy might be achieved through substitution of cadmium in vehicle parts manufacture including brake linings, tyres, paints, batteries, stainless steel etc. It is still too early to judge how effective the new European directive on 'end-of-life' vehicles (ELV) will be in reducing the proportion of cadmium used in the vehicle production process, particularly given the various exemptions contained in the directive. However, the ELV and related Restriction of Use of Hazardous Substances (RoHS) directives for electronic equipment could provide compliant BAT processes capable of substantially reducing cadmium use and environmental losses in many industrial enterprises.

## 18.3 PRODUCT SUBSTITUTION AND SUSTAINABLE DRAINAGE TECHNOLOGY

One of the criticisms of the term 'sustainable urban drainage systems' (SUDS; D'Arcy, 2013) is that the systems are sinks for persistent pollutants (if they perform as intended), thereby creating deposits of sediment contaminated by persistent pollutants such as toxic metals. Figure 18.3 shows how the stormwater management technology can move further towards sustainability if product substitution can reduce at-source concentrations of persistent pollutants in urban runoff.

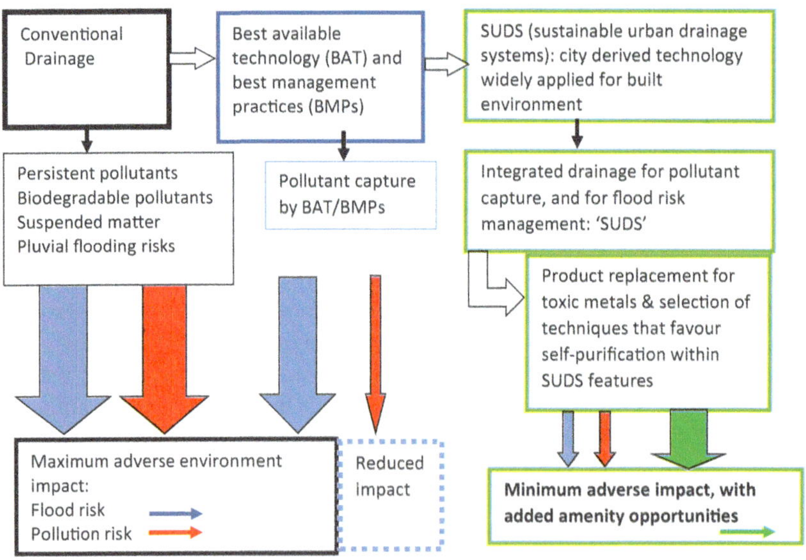


**Figure 18.3** From pipes to sustainable drainage – the need for product replacement.

Napier *et al.* (2008) reviewed road traffic as a source of pollution, and Napier *et al.* (2009) demonstrated that degradable hydrocarbons such as oil are more likely to decompose in soil/grass systems than in pond sediments. Thus a preferential use of vegetative source control techniques such as grass swales and filter strips can be advocated, rather than simply end-of-pipe ponds for managing urban runoff (Napier *et al.* 2011). For specific industrial premises processing or producing associated toxic metal or other persistent pollutants, there is scope for tailoring the design of proprietary SUDS features on the premises to maximise capture of those pollutants prior to discharge from the industrial site. The introduction of BAT in respect of flue-gas treatment and carbon adsorption scrubbers for the removal of cadmium emissions has proved successful in electroplate processing. Such approaches would be entirely consistent with the polluter-pays principle. Vehicle servicing businesses and especially tyre-related services on industrial estates for example are likely to be diffuse source hotspots. The retrofitting of catchbasin and floor trap fabric filter inserts to capture some of these materials and toxic substances has been partially successful retaining up to 20% of PAHs for example, but requires regular and costly maintenance (http://water.epa.gov; GeoSyntec/UCLA, 2005). Many roads and premises in older parts of towns and cities worldwide, are drained on a single pipe system. Reducing contamination by product substitution will be another step to increasing the scope for using sewage sludge as a resource (if free from toxic metals that can inhibit digestion processes in energy recovery plant for example, or for land disposal of sewage sludge once toxic contaminants are consistently below thresholds for application to farmland).

## 18.4 DISCUSSION

Key criteria in risk assessments of product replacement options must relate to the use and type of PP and to the efficiency of proposed mitigating strategies. Key criteria indicators include technical feasibility and efficiency, probability of achieving WFD targets, operational and investment costs, employment and supply chain impacts, impacts on drinking water production, and legislative implementation delays. Each of these indicator criteria can be allocated a score against both use and mitigating strategy to derive an overall ranking matrix for a given pollutant and its planned alternative (Lecloux *et al.* 2008).

Other toxic metals implicated in pollution from road traffic include lead, copper and zinc (Napier *et al.* 2008). Lead is a recognised priority pollutant. Lead in wheel balancing weights was not subject to the pressure for legislative control that ultimately led to alkyl lead being replaced by benzene derivative petrol anti-knock additives. That may explain why the improvement in inner-city air quality following the alkyl lead replacement was not reflected in the quality of some urban streams (it might also be a function of stream sedimentation/resuspension dynamics and year-on-year rainfall variability). Without recourse to legislation nonetheless, lead based wheel balancing weights have been replaced (in the UK and some USA States at least) by zinc products. Larger in size, the quantities of zinc metal being 'banged off' wheels onto road surfaces to be ground into dust by 40 tonnes trucks and washed into road drainage systems, is a new pollution problem awaiting detection. Zinc, whilst not as toxic as lead or cadmium, is nevertheless a significant potential pollutant and is already a widespread anthropogenic contaminant. Can higher concentrations of zinc in urban watercourses or below motorways, now be anticipated as a result of the change in wheel balancing weights? Such unintended outcomes must be anticipated as a result of product replacement strategies.

Copper has also been studied as a diffuse pollution problem in San Francisco Estuary (2007) – the replacement technology being developed by the vehicle industry in response to the California Brake Pad Partnership is ceramic units, which should result in a far smaller pollution risk. Meanwhile, the housing industry is starting to use copper as a moss-inhibiting ingredient in bitumen flat roofing materials illustrating yet another unintended hazard. Perhaps the greatest challenge is to persuade industrial product development laboratories and technologists to foresee such potential problems much sooner. This could well be at the point of product conception and design, and the most effective and sustainable substitution made at that stage rather than passing the problem to downstream users or other environmental compartments.

## 18.5 CONCLUSION

Successful substitution requires a strong evidence base, which can be costly and time consuming to achieve. It also requires long-term market confidence by industry in investment control strategies, which can be linked to acceptance of downstream users as well as to clarity regarding national policy goals and legislation related to pollution control. Regulatory decisions need to be transparent and based on proportionality, consistency and cost-benefit analysis otherwise entire products and industrial groups critical to society might be exposed to unfair or biased competition. This would only make product substitution more difficult to achieve.

For pollutants that are persistent, toxic and liable to bioaccumulation, product substitution offers the most sustainable solution to environmental pollution of the most serious kind (e.g., toxic metals and chlorinated hydrocarbons), which can adversely affect ecosystems and present significant risks to aquatic and human health. Interim actions involving capture by BAT and BMPs or SUDS offer the bottom line pragmatic interim action, along with reductions in use or losses at source.

## REFERENCES

Cefas (1994). Guideline Action Levels for the Disposal of Dredged Material. Centre for Environment Fisheries & Aquaculture Science (Cefas), Lowestoft, UK.

D'Arcy B. J. (2013). Managing Stormwater: from aspirations to routine business. gwf, S1/2013. International Issue 2013. *gwf-Wasser/Abwasser*, **154**. ISSN 0016-3651. B 5399.

E-PRTR (2008). European Pollutant Release and Transfer Register (E-PRTR). http://prtr.ec.europa.eu

European Commission (2012). Proposal for a Directive Amending the WFD and EQSD. Report COM(2011)876. European Commission (EC), Brussels, Belgium.

GeoSyntec/UCLA (2005). CIWMB Catchbasin Insert Study: Used Oil. Final Report URD3-02-005. May 2005. GeoSyntec Ltd, Portland, Oregon and UCLA, Los Angeles, California, US.

Harada M. (1995). Minamata disease: methylmercury poisoning in Japan caused by environmental pollution. *Critical Reviews in Toxicology*, **25**(1), 1–24.

Lecloux A., Eriksson E. and Ledin A. (2008). Suggestions for Reducing Emission/Exposure of PPs. Deliverable D 4.2. ScorePP, (Source Control Options for Reducing Emissions of Priority Pollutants). European Commission 5th Framework Programme, Brussels, Belgium.

Mikklelsen P. S., Lutzhoft H.-C. H., Eriksson E., Ledin A., Donner E., Scholes L., Revitt D. M., Lecloux A., Wickman T., Atanasova N., Kompare B. and Banovec P. (2008). Source Control Options for Reducing Emissions of Priority Pollutants from Urban Areas. Proc. 11th Int. Conf. on Urban Drainage (ICUD), August 2008, Edinburgh, Scotland. CD-ROM IWA Publishing Ltd., London, UK. ISBN 978 1899796 212.

Mitchell G. (2001). The Quality of Urban Stormwater in Britain and Europe. School of Geography, University of Leeds, Leeds, UK.

Napier F., D'Arcy B. J. and Jefferies C. (2008). A review of vehicle related metals and polycyclic aromatic hydrocarbons in the UK environment. *Desalination*, **226**(1–3), 143–150.

Napier F., Jefferies C., Heal K. V., Fogg P., D'Arcy B. J. and Clarke R. (2009). Evidence of traffic-related pollutant control in soil-based Sustainable Urban Drainage Systems (SUDS). *Water Science and Technology*, **60**(1), 221–30.

Napier F., Lundy L., D'Arcy B. J. and Jefferies C. (2011). Managing Hydrocarbons in the Urban Environment: Implications for BMP Selection. Proceedings of the 15th International Conference on Diffuse Pollution and Eutrophication, IWA Diffuse Pollution Specialist Group, 18–23 September 2011, Rotorua, New Zealand. OEHHA, 2013.

OEHHA (2013). Hazard Identification Documents for Butyl Benzyl Phthalate and Diisononyl Phthalate. Office of Environmental Health Hazard Assessment (OEHHA), Government of California, Sacramento, California, US.

OSPAR (2004). Cadmium. Hazardous Substance Series: Background Document on Cadmium. OSPAR Commission (Oslo & Paris Conventions; OSPAR), London, UK.

Revitt D. M., Lundy L., Eriksson E. and Viavattene C. (2013). Comparison of pollutant emission control strategies for cadmium and mercury in urban water systems using substance flow analysis. *Journal of Environmental Management*, **116**, 172–180.

Royal Commission on Environmental Pollution (1983). Lead in the Environment. 9th Report of the Royal Commission on Environmental Pollution, April 1983 (Cmnd 8852). HMSO, London, UK.

San Francisco Estuary Institute (2007). The Pulse of the Estuary: Monitoring and Managing Water Quality in the San Francisco Estuary. SFEI 532, San Francisco Estuary Institute, Oakland, CA. www.sfei.org

UKCSF (2010). A Guide to Substitution. UK Chemicals Stakeholder Forum (UKCSF), London, UK.

UKTAG (2013). Updated Recommended Data in Environmental Standards: River Basin Management (2015–21). Final report. November (2013). UK Technical Advisory Group (UKTAG) in the Water Framework Directive, London, UK.

Ziolko D., Martin O. V., Scrimshaw D. and Lester J. N. (2011). An evaluation of metal removal during wastewater treatment: the potential to achieve more stringent final effluent. *Critical Reviews in Environmental Science and Technology*, **41**(8), 733–769.

# Chapter 19

# Taking a more holistic approach to reduce diffuse industrial stormwater pollution: The Kingston Case Study (Australia)

*Tony Barrett*[1] *and Peter Morison*[2]

[1]*AECOM (E-mail: tony.barrett@aecom.com)*

[2]*Melbourne Water (E-mail: Peter.Morison@melbournewater.com.au)*

## 19.1 BACKGROUND

### 19.1.1 A diverse city

The City of Kingston is a coastal, metropolitan municipality located in the south-eastern suburbs, 20 km from Melbourne's Central Business District. The City prides itself on its beautiful beaches, world class golf courses, attractive parklands and the Ramsar-listed wetlands of Edithvale. Beyond these natural assets there is a diverse mix of land use including large residential suburbs, regional-scale shopping centres, and some of the most industrialised precincts in Australia. These precincts are home to around 5000 industrial businesses with manufacturing forming the largest business sector. Manufacturing businesses also provide the biggest employment opportunities, generating over 17,000 jobs. Kingston's manufacturing sector is also an invaluable driver of the local economy with an income of more than \$12 billion ($10^9$) annually.

### 19.1.2 A city-wide commitment to a water sensitive future

In 2012, the City of Kingston adopted the Kingston Integrated Water Cycle Strategy. This document recognised that all elements of the water cycle are linked and presented the Council's commitment to become a 'water sensitive city' by 2045. For Kingston, being a water sensitive city means water resources will be

© IWA Publishing 2017. Wealth Creation without Pollution: Designing for Industry, Ecobusiness Parks and Industrial Estates
Brian D'Arcy, Lee-Hyung Kim, Marla Maniquiz-Redillas
doi: 10.2166/9781780408330_271

managed within the boundaries of the local catchments. Its urban infrastructure will be integrated with the natural environment, providing direct health and environmental benefits. Kingston's communities will also be water sensitive and actively engaged in planning and decision making for their future.

At the heart of the strategy are a series of guiding principles and targets for each urban water stream. One of the key targets for stormwater is to achieve best practice stormwater quality objectives for all surface water runoff within the municipality by 2045. This equates to an 80% reduction in average annual loads of total suspended solids, 45% reduction in total phosphorous and a 45% reduction in total nitrogen for all surface water runoff entering waterways and coastal waters in Kingston. The baseline year against which these reductions are measured is 1999 and the scientific basis for selecting those reductions in the catchments to which they will apply can be seen in the Best Management Practice Guidelines for Urban Stormwater Management (Victorian Stormwater Committee, 1999). These reductions are commensurate with the targets of Melbourne Water, the waterways and drainage manager for the metropolitan region. Furthermore, nitrogen load reductions were set as a proportionate contribution to the 1000 tonnes target set for Port Phillip Bay as part of its Environmental Management Plan and informed by the Environmental Study conducted by the CSIRO (Harris *et al.* 1996).

To achieve these targets, Kingston City Council has developed a detailed implementation plan featuring over 50 actions. Those that are relevant to industrial precincts include:

**Action 6.3**

Provide for the treatment of the equivalent of 75 hectares of urban land to best practice stormwater management requirements till the end of the 2015/2016 financial year through the installation of raingardens, wetlands, stormwater harvesting and reuse schemes and rainwater tanks etc.

**Action 7.4**

Target the following key strategic development projects, early in the planning process, to achieve integrated water management:

- Any industrial development where new buildings or extensions exceed 500 $m^2$ (net floor area)
- Any proposed land re-zonings

**Action 7.8**

Work in partnership with EPA Victoria to deliver Council's Industry Stormwater Program (Education and Enforcement) to ensure appropriate stormwater management at industrial premises in accordance with the Local Law and state environmental legislation.

**Action 7.9**

Continue to provide support to industrial and commercial businesses on sustainable water management through the Kingston Sustainable Business Framework.

### 19.1.3 Industrial diversity

The City of Kingston's industrial precincts display an evolutionary timeline from ageing, densely populated estates in the north-west to the new and spacious industrial parks on the City's south-eastern fringe.

Each of these industrial precincts has unique characteristics that can directly influence stormwater quality or alter the risk of stormwater pollution from business activities.

The older industrial suburbs typically feature smaller buildings that offer businesses relatively inexpensive accommodation. In these suburbs, it is commonplace to find operators that have outgrown their premises and carry out polluting activities in external yard areas or even in the street.

Conversely, new industrial estates in the city's south-east capitalise on the recent provision of developable land. These estates are characterised by large industrial units set among wide, tree lined streets with vegetated grass verges (nature strips). Industrial units in these developments are the natural choice for warehousing and larger manufacturing businesses.

Despite their structural differences, these industrial estates share similar challenges in managing stormwater flow and quality from large impervious areas that are directly connected to surface water drainage and natural waterways systems (example situation in Figure 19.1).

**Figure 19.1** Surface water drainage from Kingston's industrial estates formerly flowed directly into the Mordialloc Creek without Treatment (Source: AECOM Map Portal Source: Esri, DigitalGlobe, GeoEye, i-cubed, Earthstar Geographics, CNES/Airbus DS, USDA, USGS, AEX, Getmapping, Aerogrid, IGN, IGP, swisstopo, and the GIS User Community).

### 19.1.4 Industrial stormwater quality

Stormwater runoff from these precincts enters the stormwater drainage network and discharges into the local waterways and Port Phillip Bay without treatment. This poses an obvious threat to the health of the bay as well as the recreational use of the urban beaches that make Kingston a popular destination for visitors.

In 2009, the City of Kingston conducted an extensive water quality monitoring survey to support an industrial stormwater education and enforcement programme for small to medium sized enterprises (SMEs). Melbourne University's Centre for Environmental Stress and Adaptation Research (CESAR) provided technical guidance and trialled the use of a new, cost effective stormwater monitoring technique. This innovative approach used granular activated carbon (GAC) that could be left in situ for a pre-determined number of days before being analysed for its absorbed pollutant content. In this survey, 20 small industrial sub-catchments were monitored over an 8-week period. The GAC samples were collected once a week and analysed for total petroleum hydrocarbons (TPH), copper and zinc.

The mean zinc and copper concentrations accumulated by media were 147.1 (±7.7) mg/kg and 15.3 (±0.82) mg/kg respectively (Table 19.1). Concentrations of pollutants accumulated were in the order: copper < zinc < TPH. Zinc concentrations ranged from 13 to 800 mg/kg, while copper concentrations ranged from <5 to 130 mg/kg. Hydrocarbons collected were predominately medium-weight (C10-C14; C15-C28), with occasional spikes in light (C6-C9) and heavy (C29-C36) fractions. Hydrocarbon fractions were co-correlated, so TPH (the sum of all four fractions) was generally a good summary of individual size fractions. TPH distribution was skewed by a few extremely high values, with the mean TPH concentration (810 mg/kg) approximately four times the median (212 mg/kg) (Marshall *et al.* 2008). This survey revealed unacceptable levels of pollution within Kingston's stormwater drainage network, and highlighted a complex pollution profile with significant variations in pollutant loading in each sub-catchment over time (Table 19.1).

**Table 19.1** Pollutant concentrations accumulated by GAC media in mg/kg dry weight.

| Variable | N | Median | Mean | Std Error | Minimum | Maximum | %*<DL |
|---|---|---|---|---|---|---|---|
| Cu | 264 | 11.0 | 15.3 | 0.8 | <5 | 130 | 2.7% |
| Zn | 264 | 110.0 | 147.1 | 7.7 | 13.0 | 800 | 0.0% |
| TPH C6-C9 | 264 | 13.3 | 64.0 | 13.9 | <20 | 2800 | 60.2% |
| TPH C10-C14 | 264 | 45.0 | 200.3 | 80.1 | <20 | 19,000 | 23.9% |
| TPH C15-C28 | 264 | 64.5 | 479.5 | 173.2 | <50 | 41,000 | 43.2% |
| TPH C29-C36 | 264 | 24.9 | 66.6 | 8.0 | <50 | 1300 | 77.7% |
| TPH | 264 | 212.6 | 810.4 | 254.0 | <140 | 60,135 | 15.6% |

*Proportion of samples where every hydrocarbon fraction was below the limit of detection (indicated by < and value).

### 19.1.5 The source of industrial diffuse pollution

The cause of variations in stormwater contamination at sub-catchment level becomes clear when visiting small to medium sized industrial businesses. For the vast majority of sites, there is little evidence to suggest there has ever been a significant spill or pollution incident entering the stormwater drainage system. It is however, common to find businesses conducting work activities in external yard areas. Figure 19.2 provides a good example of a yard space that poses a high risk of stormwater pollution. In this example, a business used lathes and other machines to manufacture small metal components. Most of the work was conducted inside the building; however, the business also used the external yard area for storing new materials, oil containers, old machinery and wastes. The forklift truck operating on the site regularly tracked small amounts of swarf (small metal filings) on the tyres as it moved around the site. These fine particles were contaminated with cutting oils and they could be found everywhere across the yard.

**Figure 19.2** (a) Example of a polluting external yard area with the surface water drain circled; and (b) Kingston's extensive stormwater quality monitoring survey trialled the use of GAC absorptive media (Photos: T. Barrett).

Further issues found at the site included oily residues left behind after oil drop recovery, rust stains from the metals storage area and poor storage arrangements for oils. The site operators in this instance understood the importance of correcting their oil storage facilities because they recognised the risk of an oil spill. Unfortunately, they did not consider their oil and metal residues as 'polluting' due to their seemingly insignificant appearance. This scenario of an external yard area, loaded with fine contaminants, is all too common in small to medium sized industrial premises.

## 19.2 OPPORTUNITIES FOR NEW AND REDEVELOPED INDUSTRIAL BUILDINGS

New or redeveloped industrial premises offer an excellent opportunity to tackle diffuse stormwater pollution at source. However, progress to mainstream water sensitive urban design (WSUD) and best practice approaches to site design remains frustratingly slow. This is evident in the emerging number of industrial spaces that are marketed as 'sustainable' industrial units yet feature vast roof and yard areas connected directly into the stormwater drainage system. Policies and statutes that can drive uptake of the technology are in place, but there is considerable scope for clarification and simplification. The development control process currently offers the best way of ensuring that a water sensitive design approach is taken for new developments. The Government of Victoria is currently reviewing the extension of the 'integrated water management' provisions – currently applied to residential subdivisions – to commercial and industrial development: http://www.livingvictoria.vic.gov.au/content/planning/planning-building (accessed 12.4.2015). However, while imperative for establishing WSUD as mainstream practice (Morison & Chesterfield, 2012), the regulatory amendments may take some years to materialise (Hatt *et al.* 2009).

### 19.2.1 Engaging with developers

In the absence of current mandatory requirements, greater emphasis is placed on the role of champions to encourage wider uptake of WSUD in new industrial developments.

The City of Kingston has employed an Urban and Sustainable Design Advisor to work first hand with planning applicants and developers as well as building capacity within the Council. In this role, the officer applies their architectural expertise to examine planning applications and advise applicants on best practice approaches to sustainable development. This is a broad role that considers all aspects of sustainability including energy, transport and water. For WSUD, the officer works closely with the council's Engineering Design team to determine the feasibility of proposed measures and makes recommendations for improvement.

Whilst this is not a mandatory obligation, many developers are starting to see the benefits of improved stormwater management and, by working closely with Council's Advisor, are incorporating more WSUD measures in their applications.

### 19.2.2 Best practice site layout

Previous work in Kingston investigated the optimum site layout for new industrial premises. The principal recommendation was that residential-scale WSUD measures can be used in industrial settings, providing the active work areas are 'structurally separated' from the stormwater treatment system.

Active work areas include any location used for the business activity including the storage of products and wastes. If designed and managed well, structural separation will keep rain and surface water drainage away from polluted areas, as well as provide critical on site storage in the event of a pollution incident. Typical approaches to structural separation include roofing, canopies, covered storage areas, cut off trenches and drains that are connected to blind sumps or the sewer network under a trade waste agreement where appropriate (Figure 19.3).

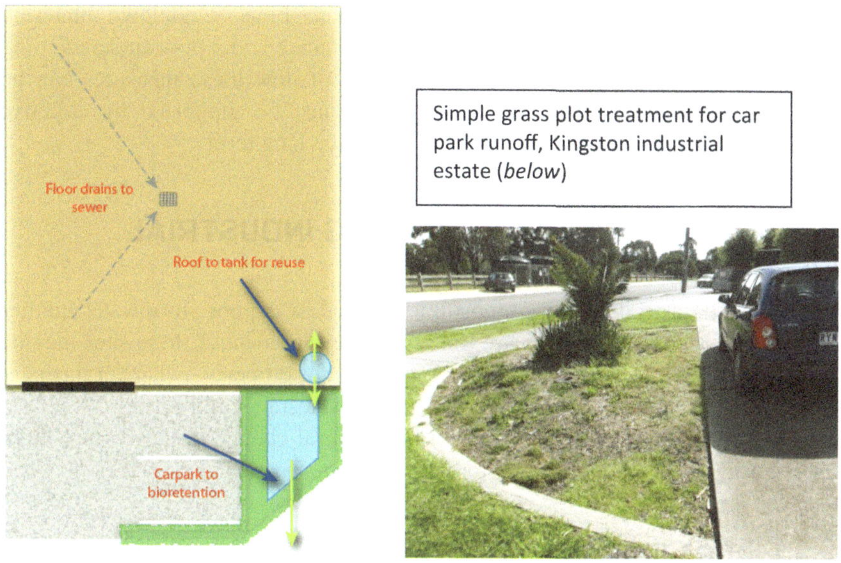


**Figure 19.3** An example of best practice site layout using structural separation of active work areas (Wong & Walsh, 2006).

With 'dirty' areas excluded from the stormwater pathway, runoff from the industrial site will have a similar pollutant profile as typical urban development. This means standard WSUD measures can be selected to treat the runoff from the roof and car parking areas (Wong & Walsh, 2006). Further design modifications can be made to enhance natural surveillance at a site to reduce pollution from deliberate or neglectful site activities. Such measures include building orientation, locating buildings to the rear of a site to omit 'out of sight' yard spaces and replacing subsurface drains with vegetated stormwater channels to make it easier to trace pollution incidents to particular site (Kay, 2006).

These recommendations for best practice site design were adopted by the City of Kingston as a new benchmark for industrial development. Using these findings as an evidence base, Kingston's Urban and Sustainable Design Advisor was able to encourage more effective stormwater management measures for new industrial development applications.

In the absence of regulatory drivers, implementation of WSUD in new industrial development is heavily reliant on the Council's ability to engage with developers at the earliest possible stage in the planning process. This is a great challenge and one that most councils are unlikely to resolve in every planning application. For the time being, sites constructed without stormwater quality controls will inevitably form the majority of new industrial developments.

Despite this outcome, it is clear that structural separation measures alone cannot reduce diffuse stormwater contamination. Fine, windblown material, liquids tracked on vehicle tyres, a lack of maintenance and a poor understanding of operational impacts are just a few examples of how these measures can be compromised. Structural separation measures must be supported by effective operational controls if they are to be effective in the long term.

## 19.3 OPPORTUNITIES AT EXISTING INDUSTRIAL PRECINCTS

For existing industrial estates, opportunities to address poor stormwater quality runoff with structural measures are limited. Redevelopment of these sites may be several decades away and there are few incentives for property owners or business operators to invest in site modifications. For these estates, retrofitting WSUD at street and district scales combined with behavioural change initiatives are more effective ways to tackle industrial stormwater contamination.

### 19.3.1 Engaging with business operators

Industrial small to medium sized enterprises are largely unregulated by the Environmental Protection Authority (EPA) Victoria. Where EPA intervention does occur, it is most often due to complaints of poor environmental practices or in response to a reported incident. While effectively tackling the more significant, often visual issues, reactive regulation does little to reduce the persistent levels of diffuse pollution identified in Kingston's water quality monitoring surveys.

In 2008, the City of Kingston completed an extensive research project that examined the role of Local Government in reducing diffuse industrial pollution. The project, funded by the Commonwealth Government's Coastal Catchments Initiative (CCI) investigated the causes of stormwater contamination at site level. It also trialled the effectiveness of education and enforcement campaigns for small to medium sized industrial businesses. With the agreement of EPA Victoria, the Council developed a simple enforcement model that would give the site operator the opportunity to address any minor contamination issues without penalty. Where significant issues were identified, or if the business repeatedly failed to address the minor issues, EPA Victoria would be notified to take further action.

The Council's Industry Stormwater Officers visited around 140 businesses in the initial trial. Of these, 47 were found to be operating in a manner that directly contributed to, or presented an unnecessary risk of stormwater contamination. Each site operator was given a set number of days to address the issues, based on the significance of the risk. Nearly half of the businesses resolved the issues before the first follow up visit, whereas only two businesses failed to make any changes to their activity. These businesses were passed to EPA Victoria for further action. The audit pilot programme successfully demonstrated that local government organisations can improve poor work practices at industrial SMEs to help prevent stormwater contamination. Table 19.2 presents the number of visits taken before site improvements were made.

**Table 19.2** Number of follow up visits required to resolve identified stormwater pollution issues at 47 businesses.

| | Completed by 1st Follow Up Visit | Completed by 2nd Follow Up Visit | Completed by 3rd Follow Up Visit | Completed by 4th Follow Up Visit |
|---|---|---|---|---|
| No. of Businesses | 22 | 15 | 8 | 2 |

Perhaps more interestingly, the programme revealed a widespread lack of understanding where business managers were unaware of the impacts their activity had on the local environment. This study had a marked effect on the future of Kingston's work in industrial stormwater quality management. The Council understood that, even with best practice site design, a poorly managed business could still pollute a local waterway, or even compromise the performance of local WSUD schemes. Kingston City Council made a commitment to employ a full-time Industry Stormwater Officer on the back of the study. This was to continue the successful programme as well as educate businesses that were to benefit from street-scale industrial retrofit schemes on the importance of getting it right on site (see Figure 19.4).

**Figure 19.4** (a) A combination of permeable paving in parking areas (*left*) and porous block paving for the road (*right*) constructed in the Mordialloc Industrial Estate; (b) a business manager demonstrates some of the operational and site modifications made during the programme; and (c) receiving a certificate of good practice. (Photos: T. Barrett).

### 19.3.2 Retrofitting for change

In Victoria, local government organisations have a responsibility to manage minor road networks including the local street and property drainage systems that feed into the larger, regional drains, rivers and creeks (ss. 198, 201 Local Government Act 1989).

Like most councils in Victoria, the City of Kingston delivers a capital programme for road and drain reconstruction projects. These projects provide an excellent opportunity to retrofit WSUD at street and district scales (Figure 19.4). Recognising the multiple benefits and cost efficiencies associated with WSUD retrofit schemes, the City of Kingston recently applied their long-established residential experience to retrofit an existing industrial estate in Mordialloc with WSUD, during a road reconstruction project (West, 2013).

The scheme included a redesign of three industrial roads to treat and harvest over 4 mega litres of stormwater from the roads and factory roofs within the estate. This is the first time an existing industrial estate has been retrofitted with WSUD in Australia.

Stormwater runoff from the site is treated through a combination of custom built and proprietary products including litter traps in side-entry pits, gross pollutant traps, porous block paving and permeable paving. A large bioretention swale provides the final level of treatment before being pumped to a 240 kl above-ground storage tank where it is used to irrigate the adjacent cricket pitch. Kingston City Council also access the water for their street tree watering programme, ensuring newly planted street tress in the vicinity thrive through the summer season.

Without this system, untreated stormwater runoff from the industrial estate would enter the sensitive estuarine reaches of the Mordialloc Creek. Furthermore, the cricket pitch and street tree watering programme would rely on high quality mains water, adding to increased water stress during the summer months. The benefits of this district-scale stormwater harvesting and reuse scheme extend far beyond an improved urban water cycle and waterway health. It also provides the community with a safe playing surface through the summer, the street trees help cool local streets, the scheme educates the local community on using water that is fit for purpose and is a powerful demonstration of community leadership. In an effort to improve the scheme's resilience, Kingston's Industry Stormwater Officer visited every business in the catchment. The key objective for these visits was to educate the site operators on why their street is different and the reasons why the scheme was constructed. Whilst relatively new, the outcomes for this project are already informing decisions on future stormwater projects for Kingston and other Councils across Australia.

## REFERENCES

City of Kingston, AECOM (2012). Kingston Integrated Water Cycle Strategy. Availability: http://www.kingston.vic.gov.au/Explore-Kingston/Environment/Sustainability-in-Council

EPA (2012). Water related legislation. Online: http://www.epa.vic.gov.au/about_us/legislation/water.asp#sepp_waters (last updated 21 February 2012).

Harris G., Batley G., Fox D., Hall D., Jernakoff P., Molloy R., Murray A., Newell B., Parslow J., Skyring G. and Walker S. (1996). Port Phillip Bay Environmental Study Final Report. CSIRO, Collingwood.

Hatt B., Morison P., Fletcher T. and Deletic A. (2009). Adoption Guidelines for Stormwater Biofiltration Systems. Facility for Advancing Water Biofiltration, Monash University, June 2009. http://www.monash.edu.au/fawb/products/index.html

Kay E. (2006). Industry planning Controls: Initial Investigation, Part of the WSUD in Industrial Sites and Precincts Project. Environment and Land Management. Prepared for the City of Kingston, March 2006.

Marshall S., Barrett T. and Pettigrove V. J. (2008). Use of Absorptive Media to Monitor Stormwater Contamination in Small Urban Drains. In: Proceedings of the Highways & Urbanisation Symposium, June 2008. Springer Press, Madrid, Spain.

Morison P. J. and Chesterfield C. (2012). Enhancing the management of urban stormwater in a new paradigm [online]. In: WSUD 2012: Water Sensitive Urban Design; Building the Water Sensitive Community; 7th International Conference on Water Sensitive Urban Design, 21–23 February 2012, Melbourne Cricket Ground. Engineers Australia, Barton, A.C.T., pp. 910–917, ISBN: 9780858258952. Availability: http://search.informit.com.au/documentSummary;dn=829417691355605;res=IELENG

Morison P. J., Brown R. R. and Cocklin C. (2010). Transitioning to a waterways city: municipal context, capacity and commitment. *Water Science & Technology*, **62**(1), 162–171.

Victorian Stormwater Committee (1999). Urban Stormwater Best-Practice Environmental Management Guidelines. [online] CSIRO Publishing, Port Melbourne. Availability: http://www.publish.csiro.au/issue/3822.htm

West A. (2013). Mordialloc Industrial Precinct Case Study, Presented by Clearwater. Online: http://www.clearwater.asn.au/resource-library/case-studies/mordialloc-industrial-precinct.php

Wong T. and Walsh G. (2006). Establishing Water Sensitive Urban Design in Industrial Sites and Precincts, Summary Report, Ecological Engineering. Prepared for the City of Kingston, February 2006.

# Chapter 20

# Beyond legislation – working together to protect the water environment

*Phil Chatfield*[1] *and Lian Lundy*[2]

[1]*Welsh Government, Cathays Park, Cardiff CF10 3NQ, UK (E-mail: phil.chatfield@wales.gsi.gov.uk)*

[2]*Middlesex University, The Burroughs, Hendon, London, NW4 4BT, UK (E-mail: L.Lundy@mdx.ac.uk)*

## 20.1 INTRODUCTION

Although legislation on the protection of the water environment has improved in many countries in recent decades, there are practical limits to the application of a legislative approach and it is often difficult to enforce. In many cases it fails to prevent environmental harm, because when water pollution occurs, it is the result of ignorance and neglect rather than deliberate acts. Over the last 20 years a range of alternative, often voluntary, measures have been developed to prevent pollution and protect the water environment. The most successful of these initiatives go beyond legislation, with regulators and other stakeholders, including the relevant industrial sectors, working together to promote good practice and improved standards. This paper briefly outlines the legislative framework which provides the context for these voluntary initiatives, defines terminologies, describes the benefits for regulators, industry and the environment of working together and provides case studies from the European Union (EU).

© IWA Publishing 2017. Wealth Creation without Pollution: Designing for Industry, Ecobusiness Parks and Industrial Estates
Brian D'Arcy, Lee-Hyung Kim, Marla Maniquiz-Redillas
doi: 10.2166/9781780408330_283

## 20.2 WHY WORK TOGETHER?

Whilst data on which to quantify the benefits of collaborative working is scarce (Slater *et al.* 2007; Reed, 2008; De Stefano, 2010), there is overwhelming support for the approach within a raft of European legislation (see Box 1) and across a broad range of sectors, for example health care, education and transport planning (Carnwell & Carson, 2005; Genç & Öykü Iyigün, 2011; Frantzeskaki *et al.* 2013). Many terms are widely used, often inter-changeably, to describe such collaborative efforts (e.g., partnership working, stakeholder consultation, public participation etc.) and an overview of commonly used terminologies is given in Table 20.1.

**Table 20.1** Overview of descriptors commonly used to describe various aspects of collaborative working.

| Descriptors | Term | Definition |
|---|---|---|
| Common descriptors for groupings | Stakeholders | Individuals/organisations who have something at stake, e.g., are impacted by or have influence on the decision being taken |
| | Collaboration | A bond, union or partnership, characterised by mutual goals and commitments (Carnwell & Carson, 2005) |
| | Partnership | A shared commitment, where all partners have a right and an obligation to participate and will be affected equally by the benefits and disadvantages arising from the partnership (Carnwell & Carson, 2005) |
| | Learning Alliance | A group of individuals or organisations with a shared interest in innovation and the scaling up of innovation in a topic of mutual interest (Batchelor & Butterworth, 2008) |
| Common descriptors for processes | Stakeholder consultation | Process of providing a forum for stakeholders to speak up and voice opinions only. Those opinions then may or may not be taken into account fully by decision makers (World Bank, undated) |
| | Public participation | Allowing people to influence the outcome of plans and working processes (CIS, 2003) |
| | Partnership working | A spectrum, ranging from informal networking forums, consulting and sharing information and intelligence, through to formal strategic alliances where partners come together to achieve common goals by changing the way that they work (Rocket Science, 2006) |
| | Social learning | Capacity building of individuals and organisations; seen as an alternative complimentary policy instrument in governance. Increases ability to handle uncertainty and change (van Herk *et al.* 2011) |

With a focus on industry-regulator partnerships, the literature identifies a range of reasons for collaborative working. The development of a forum where industry and regulators can work together provides a constructive opportunity for those affected by decisions (e.g., industry) to influence decisions being taken by regulators which may affect their activities. It can facilitate the breakdown of legislative and institutional barriers to implementing changes and provide a more consistent, comprehensive approach which makes the best use of available resources in a co-ordinated and cost-effective way (Common Implementation Strategy [CIS], 2013). It can support the development and consideration of novel and innovative options which may be implemented with greater ease, ensuring that decisions taken are workable and acceptable within national and local regulatory and operating contexts (van Herk *et al.* 2011). Collaborative working can raise organisational awareness of environmental issues, make use of the knowledge and expertise held by a wider range of stakeholders and thus improve the quality and creativity of measures adopted. It can also generate approaches which have higher levels of regulator/organisational/sectoral (and wider public) acceptance, commitment and support, decreasing levels of litigation, and misunderstandings associated with emerging plans and initiatives. In turn this can lead to fewer operational delays and more effective implementation of agreed measures (CIS, 2003).

In an evaluation of why partnerships emerge, Benson *et al.* (2013) state that rational organisations will opt to work together if the benefits of doing so outweigh costs of collaborating (where costs can be financial or relate to level of time input required). Similarly, Slater *et al.* (2007) see the development of partnership working as a strategic and operational approach to addressing the increasingly stringent requirements faced by many sectors in many fields; that through working together organisations and regulators combine resources and can achieve objectives that they could not achieve individually.

**BOX 20.1: LEGISLATIVE DRIVERS SUPPORTING COLLABORATIVE WORKING BETWEEN ORGANISATIONS AND REGULATORS**

The 1998 UNECE Aarhus Convention on access to information and public participation clearly states that sustainable development can only be achieved through the involvement of all stakeholders (UNECE, 1998). The EU Water Framework Directive (WFD, 2000) states its success relies on co-operation and coherent action at European, national and local levels, as well as involvement of the public. Article 14 of the WFD requires Member States (MS) to 'encourage the active involvement of all interested parties ... in the production, review and updating of the river basin management plans' (EU WFD, 2000). The EU Floods Directive (2007) contains almost identical language with regard to the production, review and updating of flood risk management plans. The Integrated Pollution Prevention and Control (IPPC) Directive (2008) requires that the public are given effective opportunities to participate in permitting decisions, and public participation forms a key component of the Strategic Environmental Impact Assessment Directive (SEA, 2001).

## 20.3 BUILDING AND WORKING IN PARTNERSHIP

There is no single correct way to undertake collaborative working (CIS, 2003). Various forms of partnership can and do work in a number of ways; specific 'best practice' for collaborative working has yet to be agreed and there is no single 'blue print' which can be followed to ensure successful outcomes are achieved (De Stefano, 2010). However, various authors have identified a series of generic principles or elements of good practice in collaborative working which are reported to facilitate successful partnerships (CIS, 2003; Reed *et al.* 2008; van Herk *et al.* 2011) which could be applied to the development of industry-regulator partnerships. These include:

- Partnership working should be considered by all parties as early as possible and continue throughout any environmental decision-making process
- Key stakeholders need to be identified, analysed and represented systematically
- The role of each partner is identified and clear to all partners
- The aim and objectives of the partnership are agreed among all stakeholders at the beginning of the process
- Methods for collaborative working and communication should be tailored to the specific context
- Highly skilled facilitation is essential to enable the development of a supportive atmosphere where suggestions, ideas and conflicts are addressed in a timely manner
- Local and scientific knowledge should be integrated

In a study of collaborative working practices between industry and regulators aimed at tackling diffuse industrial pollution, Usman (2001) identified a number of important factors as influencing the success of partnership activities. These included:

- Awareness of the causes of diffuse pollution amongst industrial partners
- Completion of detailed research to understand target audiences' needs and concerns
- Development of support materials targeted appropriately to the audience
- A focus on practical solutions
- A good relationship and trust between regulators and potential partners

Successful partnerships grow incrementally and evolve through the building of trust and shared experiences, both time consuming concepts (Slater *et al.* 2007). Key individuals – or local champions – are critical in the early stages of partnership building to both bring on board other partners and strengthen commitment to the process (Morris, 2006). As in many arenas of life, trust is hard won but easily lost; openness, good communication, equivalent status between partners, shared experience and transparency are all crucial components in building the patience and mutual trust required to enable new ways of working to be developed and implemented.

Based on the above discussions, Table 20.2 sets out the key stages in building a successful partnership with supporting descriptors describing their application in relation to an industry-regulator example.

**Table 20.2** Key stages in building a successful partnership with descriptors elaborating their application in relation to an industry-regulator example.

| Key Stages in Partnership Building | As Applied to a Hypothetical Industry-Regulator Example |
|---|---|
| Establish the decision-making context | How to reduce diffuse pollution loadings from an industrial estate entering an adjacent watercourse |
| Undertake a stakeholder analysis to identify key stakeholders | Representatives of each company within the industrial estate, the regulator and municipality |
| Invite key stakeholders to a meeting and evaluate the need for further stakeholder involvement | Hold on-site and include a field visit to discuss/observe issues directly |
| As a partnership collaboratively identify aims and objectives for the partnership | Reduce diffuse pollution by reviewing and upgrading on-site 'house keeping', waste disposal and management of surface runoff measures |
| Ensure partners are committed, willing and have required resources | Support partnership members to cost out any revisions and savings/sources of funding to implement changes |
| Develop partnership terms of reference | Clarify the roles and remits of all stakeholders, working structures and practices including arrangements for sharing information and monitoring and review of activities undertaken |
| Start work! | Establish and agree a time line for proposed activities |
| Review and evaluate | Develop and implement an annual review of activities undertaken |

## 20.4 CASE STUDY EXAMPLES

### 20.4.1 Oil Care Campaign

Against a background of a steady increase in the number of oil related pollution incidents in England and Wales in the early 1990s, the environmental regulator (the National Rivers Authority – from 1996 the Environment Agency) worked with representatives of the oil industry to address the problem. Working together they developed a shared understanding of the causes of these pollution incidents and identified simple measures to address these, such as improved delivery and handling procedures and secondary containment for oil storage containers. These partners sought others with a common interest, for example local authorities, motor oil retailers and oil delivery companies, to develop a joint campaign designed to address the principle causes of oil pollution with clearly identified target audiences and appropriately targeted materials (Oil Care Campaign, 2015).

The need for an independent identity for the campaign was recognised at an early stage, resulting in the development of a title – The Oil Care Campaign – and a logo (see Figure 20.1). A strong brand, independent of the individual partners, resulted in widespread acceptance and use across the UK and emphasised the partnership basis for the campaign. Following the launch of the campaign, growth in the number of oil pollution incidents was reversed and numbers stabilised. A shared recognition of the limitations of this voluntary approach was also successful in encouraging Government to legislate to put in place minimum standards for oil storage and secondary containment (Oil Storage Regulations for England 2001, Scotland 2006, Northern Ireland 2010, and Wales in 2015).

**Figure 20.1** Oil Care Campaign logo.

## 20.4.2 Pollution Prevention Pays

Pollution Prevention Pays is a long-running initiative to target industry, in particular small to medium sized enterprises, with practical advice and guidance on the prevention of pollution. The creation of a single environmental regulator for the water environment in England and Wales in 1989, the National Rivers Authority (NRA), brought together staff with a wide range of experience in dealing with pollution problems. The creation of a single, national report on pollution incidents (NRA, 1993) provided a data platform for the development of a proactive approach to preventing pollution. Simple guidance, initially developed in the Thames Region, provided practical advice on how pollution could be prevented, for example through:

- improved handling and storage of chemicals and fuels,
- better understanding of drainage systems and their use and maintenance,
- the development of emergency response plans.

These Pollution Prevention Guidance (PPGs) documents were initially developed based on the experience of the regulator's field staff, with limited input from the target audience.

In order to increase the effectiveness of this guidance, an internal survey of field staff was undertaken. This indicated that there was a perception that the majority of incidents were the result of the activities of small to medium sized enterprises, for whom, because of limited resources, environmental issues were a low priority. A working group on pollution prevention in the NRA considered that a short video, with supporting materials such as workplace posters and a summary leaflet, linked to the PPGs, would be effective in addressing the problem.

A script was developed by a media company in consultation with NRA staff and taking account of research with representatives of the target audience. The video explained the common causes of water pollution and offered simple and inexpensive solutions. Responses from companies receiving the video were very positive, with many making changes to their site and procedures and often using it in staff training.

The video was refreshed in 1996 with the creation of the Environment Agency for England and Wales and the Scottish Environment Protection Agency, who also chose to use it. Since then, the contents have been regularly reviewed to take account of changes in legislation, and new pollution prevention resources.

In 2004 the whole Pollution Prevention Pays pack was again substantially reviewed and refreshed. The summary leaflet became a more detailed booklet which includes checklists to encourage the user to take action to prevent pollution and directs them to the PPGs for further information. A new script was developed for the video, which was re-filmed and produced as a DVD. New workplace posters, designed to reinforce key messages from the booklet and DVD were also produced. These revisions of content were undertaken in full consultation with industry and have been supported by trade bodies such as the Federation of Small Businesses. Further advances in communications technology have made it possible to make the Pollution Prevention Pays programme available on the internet on You-tube (e.g., see http://www.youtube.com/watch?v=LyDjC-KXANI&feature=youtu.be).

More than 20 years since its original development, the concept of improving the environment by providing readily accessible, simple advice and guidance is still relevant. The Pollution Prevention Pays resources remain effective as a result of engagement with the target audience, regular reviews and a consistent message. Experience shows that pollution prevention work is never finished. New companies are continually being formed, staff change and lessons learned need to be reinforced. The PPGs and Pollution Prevention Pays have become established as vital resources for smaller companies. In addition, they have been extensively used by larger companies with established environmental management systems as training tools, to share good practice and to provide consistency both within and between companies.

### 20.4.3 The Voluntary Initiative (reducing the environmental impacts of pesticides)

Following a request from the UK government to consider alternatives to the introduction of a legislative pesticide tax, the UK Crop Protection Association together with a range of stakeholder organisations developed the Voluntary Initiative (VI), launched in 2001. The central aim of the VI was to minimise the environmental impact of pesticides and a range of voluntary actions aimed at addressing the issues from a variety of perspectives were developed. These include:

- The use of Crop Protection Management Plans – self-assessment forms to support farmers in assessing the risks associated with crop protection on their farm
- Development of a National Sprayer Testing Scheme (NSTS) – a programme to ensure spraying equipment is correctly maintained and applies pesticide without spillage
- Development of a National Register of Sprayer Operators (NRSO) – provides training and information to sprayers to promote their continued professional development
- Implementation of the Water Catchment Protection Project – a collaboration between the farming, crop protection and water industries to 'show case' success within the six case study catchment areas and promote dissemination of best practices on a national scale
- Indicator Farms Project – established to provide case study data on the environmental impacts associated with implementation of the VI.

Communication of information to all stakeholders at the water catchment level was central to the VI resulting in the generation of information materials in a variety of formats. The key message being communicated is that pesticides are a problem, that part of the problem is caused by agricultural practices but that there are simple inexpensive solutions to addressing the issues. Information is circulated and shared through the distribution of free weekly advice by email or text messages, local newsletters providing updates on project progress and local meetings on specific topics where attendees gain new knowledge or skills. A particular emphasis is placed on the communication of information on the impacts of wet weather and associated ground conditions on the quality of receiving waters. Within this context, text messaging farmers, sprayers and agronomists on current ground conditions and the impact of spraying has been well received. This initiative is seen as being strongly linked to achieving compliance with a range of European requirements, from the EU Drinking Water Directive to the EU Water Framework Directive and the World Health Organisation (WHO) guidelines for drinking water.

On completion of the 5-year programme, all of the Government-set targets were achieved or exceeded, with experience in various catchments suggesting that it takes approximately 15 months for positive results to be seen. In relation to water quality, monitoring data indicates that the use of farmer-led initiatives can lead to an improvement in water quality of over 90%. For example, in one catchment area the number of days in which the water quality standard for pesticides was exceeded was reduced by 98%. However, this level of success was not achieved in all areas with little progress in reducing total pesticides reported for one catchment and only reductions in one particular pesticide reported in another. The reasons behind this differential success are not clear but factors such as the degree of local stakeholder engagement, weather pattern and soil management practices are put forward as highly influential factors. In relation to changing practices, over 80% of arable land is now sprayed using NSTS tested equipment operated by NRSO members (VI Annual Report, 2007). Crop Protection Management plans have been returned for 36% of arable land (although the report states that the actual number of these forms completed and not returned is thought to be in excess of this figure) with an increasing number of agronomists gaining Biodiversity and Environmental Training for Advisors (BETA) qualifications.

### 20.4.4 The Graphics business sector, Denmark

As part of its guidance document on public participation in relation to the WFD (CIS, 2003), a range of case studies which give an overview of approaches to public participation in the field of water management within the EU are presented both to give an overview of public participation approaches and to provide motivation to potential public participation partners. A pertinent example of a successful partnership between industry and regulators is that relating to an initiative between the Graphics Business Sector Association, graphics companies and the Danish Environmental Protection Agency (EPA), aimed at reducing water consumption within the graphics sector (CIS, 2013). The partnership's key activities focused around selected companies undertaking demonstration actives (involving testing the use of new equipment) and the sharing of new knowledge and results generated. Demonstration equipment was purchased using project finances, with partners providing co-financing in terms of allocation of staff time to work on demonstration activities.

The partnership is reported to have successfully addressed several challenges commonly identified as barriers to changing practice including working collaboratively with competitors (e.g., information on environmental problems and related improvement opportunities disseminated through business and Danish EPA networks), lack of interest/motivation of companies to improve environmental performance (i.e., awareness raising of potential incentives) and lack of access to financing (e.g., combination of project and co-financing to buy and test new equipment). Benefits cited include a 70–90% reduction in water consumption and the development of a positive attitude by a business sector towards implementing cleaner practices. Further benefits include the generation of robust data on feasibility and incentive measures that policymakers can use in the further development of environmental regulations, supporting the development of regulation that is workable within the sector's current operating parameters. The initial project funding taken from the government budget was later saved in terms of reduced costs for wastewater treatment plant operators.

## 20.5 CONCLUSIONS

The benefits of a pro-active approach to preventing diffuse pollution from industrial developments, in which regulators and operators work together, may be difficult to quantify. However, a number of successful examples from the EU demonstrate that, where objectives can be agreed and trust is built, this type of partnership working can have important benefits, including the protection of companies' reputation, reduced costs in regulation and enforcement, and an improved local environment. Experience has shown the importance of operators understanding the causes and impacts of diffuse pollution, careful preparation in the development of any initiative, and appropriately targeted and simple, consistent guidance on good practice and practical solutions. Because industrial estates tend to have regular changes in tenants and activities, it is important to maintain and refresh these types of initiative if the benefits are to be maintained.

## REFERENCES

Benson D., Jordan A., Cook H. and Smith L. (2013). Collaborative environmental governance: are watershed partnerships swimming or are they sinking? *Land Use Policy*, **30**, 748–757.

Carnwell R. and Carson A. (2005). Understanding partnerships and collaboration. In: Effective Practice in Health and Social Care: A Partnership Approach, R. Carwell and J. Buchannan (eds), Open University Press, Maiden Head.

CIS (2003). Common Implementation Strategy for the Water Framework Directive (2000/60/EC). Guidance document no. 8 Public Participation in relation to the Water Framework Directive.

Frantzeskaki N., Wittmayer J. and Loorbach D. (2013). The role of partnerships in 'realising' urban sustainability in Rotterdam's City Ports Area, The Netherlands. *Journal of Cleaner Production*, **65**, 406–417.

Genç N. and Öykü Iyigün N. (2011). The role of organizational learning and knowledge transfer in building strategic alliances: a case study. *Procedia Social and Behavioral Sciences*, **24**, 1124–1133.

NRA (1993). Water Pollution Incidents in England and Wales. National Rivers Authority. ISBN 1 873160 550.

Oil Care Campaign (2015). Oil care. http://www.oilcare.org.uk/ (accessed 27 March 2017)

Reed M. S. (2008). Stakeholder participation for environmental management: a literature review. *Biological Conservation*, **141**, 2417–2431.

SEA (2001). Directive on Strategic Environmental Impact Assessment (Directive 2001/42/EC). Available at http://ec.europa.eu/environment/eia/sea-legalcontext.htm

Slater R., Frederickson C., Thomas C., Wield D. and Potter S. (2007). A critical evaluation of partnerships in municipal waste management in England. *Resources, Conservation and Recycling*, **51**, 643–664.

UNECE (1998). Convention on Access to Information and Public Participation in Decision-making and Access to Environmental Matters. Available at: http://www.unece.org/fileadmin/DAM/env/pp/documents/cep43e.pdf

Usman F. (2001). Designing Effective Pollution Prevention Campaigns, Environment Agency Research and Development Technical Report P434, March.

van Herk S., Zevenbergen C., Ashley R. and Rijke J. (2011). Learning and action alliances for the integration of flood risk management into urban planning: a new framework from empirical evidence from The Netherlands. *Environmental Science and Policy*, **14**, 543–554.

VI Annual Report (2007). Annual report 2006–2007. Available from: http://www.voluntaryinitiative.org.uk/about/annual-reports/ (accessed 27 March 2017)

World Bank (2004). Stakeholder Consultation and Participation in MSWM Planning http://siteresources.worldbank.org/INTUSWM/Resources/463617-1202332338898/socialassesstoolkit.pdf (accessed 21 March 2017)

# Index

IWA Publishing's authorised EU representative for General Product Safety Regulations is Diane D'Arras, 15 rue Duret, 75116 Paris, France, e-mail: safety@iwap.co.uk.

Printed and bound by CPI Group (UK) Ltd, Croydon, CR0 4YY
16/06/2026
02140408-0001